Health Informatics

Practical Guide for Healthcare and Information Technology Professionals

Sixth Edition Supplement

ROBERT E. HOYT MD FACP

Editor

ANN K. YOSHIHASHI MD FACE

Associate Editor

Health Informatics

Practical Guide for Healthcare and Information Technology Professionals

Sixth Edition Supplement

Disclaimer

Additional Resources

Visit the textbook companion website at http://informaticseducation.org/

Editors

Robert E. Hoyt MD, FACP

Director Health Informatics

College of Health

University of West Florida

Pensacola, FL

Ann K. Yoshihashi MD, FACE

Guest Lecturer

College of Health

University of West Florida

Pensacola, FL

Contributors

Alison Fields BS
Master of Public Health Program
College of Health
University of West Florida
Pensacola, Florida

Robert Hoyt MD FACP
Director, Health Informatics
College of Health
University of West Florida
Pensacola, Florida

Harold Lehmann MD PhD
Director, Division of Health Sciences Informatics
Professor of Health Sciences Informatics
Johns Hopkins University School of Medicine
Baltimore, Maryland

Steve Magare MS
Research Officer, Health Informatics
KEMRI-Wellcome Trust Programme
Nairobi, Kenya

Sarita Mantravadi PhD, MS, MPH, CPH, CHES
Clinical Assistant Professor
College of Health
University of West Florida
Pensacola, Florida

Naomi Muinga MS
Research Officer, Health Informatics
KEMRI-Wellcome Trust Programme
Nairobi, Kenya

Gleber Nelson Marques PhD
Professor, Computer Science
Mato Grasso State University
Mato Grasso, Brazil

Chris Paton BMBS, BMedSci, MBA, FACHI
Group Head for Global Health Informatics at the Centre for Tropical Medicine
University of Oxford
Oxford, England
Administrator, Health Informatics Forum
www.healthinformaticsforum.com
Director of New Media Medicine Ltd
www.newmediamedicine.com

Dallas Snider PhD
Assistant Professor, Computer Science
Hal Marcus College of Science and Engineering
University of West Florida
Pensacola, Florida

Preface to the Sixth Edition Supplement

Health Informatics is a very dynamic information science, unlike most of the physical sciences. This means that content needs to be continuously updated and scrutinized. Authors and editors should be on the look out for new topics as they appear on the horizon. Because we utilize self-publishing it is far easier to produce new chapters that discuss emerging topics.

This supplement includes three new chapters we hope are of interest to our current and future readers. As we look down the road to a seventh edition, we anticipate that these three chapters will be part of a new two volume textbook.

Chapter 23 presents the *Introduction to Data Science*, which is an exciting new field that should be of interest to all informaticists. Topics such as Big Data and machine learning algorithms are hot topics in the lay press. We brought together a computer scientist and a statistics expert to try to combine information from these two fields. Traditionally, machine learning (computer science) and statistical modeling (math and statistics) are taught separately, even though they share a great deal in common. Isn't it time these two fields talk to each other?

In 2015 I passed the Board Exam for Clinical Informatics. In the process of studying for the exam it became clear that more information about clinical decision support was needed and hence the reason for chapter 24 *Clinical Decision Support*. Dr. Harold Lehmann from Johns Hopkins and I present what we currently know about clinical decision support, to include our usual list of recommended resources and known challenges.

It has become apparent that there is an international hunger for more information about Health Informatics. We continue to have requests for textbook downloads from every continent. This includes a few surprises from countries, such as Iran, Saudia Arabia and several small African countries. We have supported informatics initiatives with our grants in those same areas. This prompted us to create the chapter 25 on *International Informatics*. We were fortunate to find authors from several countries to provide content.

Like our other chapters, we tried to include case studies that highlight interesting national and international initiatives. We have made every attempt to provide the most up-to-date information about health informatics recent information and the most interesting concepts. We are dedicated to presenting the issues fairly and objectively and have avoided the hype some times associated with new technologies. It is also a resource/reference for people in the field, reviewing for clinical informatics board and for both graduate and undergraduate courses.

One of the goals of this book is to promote and disseminate innovations that might help healthcare workers as well as technology developers. The fact that we mention specific hardware or software or web-based applications does not mean we endorse the vendor; instead, it is our attempt to highlight an interesting concept or innovation that might lead others in a new direction.

We appreciate feedback regarding how to make this supplement as user friendly, accurate, up-to-date and educational as possible. Please note that book proceeds will be donated to support the advancement of health informatics education.

Robert E. Hoyt MD FACP

Diplomate, Clinical Informatics

Ann K. Yoshihashi MD FACE

Acknowledgements: We would like to acknowledge Alison Fields for her invaluable support in performing literature searches, formatting and proofreading that were critical in creating the supplement.

Table of Contents

Chapter 23

Introduction to Data Science

ROBERT HOYT

DALLAS SNIDER

S. MANTRAVADI

Learning Objectives

After reading this chapter the reader should be able to:

- Define the field of data science
- Enumerate the general requirements for data science expertise
- Differentiate between modeling using statistics and machine learning
- Describe the general steps from data wrangling to data presentation
- Discuss the characteristics of big data
- Discuss how data warehousing and relational database systems are important in data science
- Provide examples of how data analytics is assisting healthcare
- Discuss the challenges facing healthcare data analytics

"The ability to take data, to be able to understand it, to process it, to extract value from it, to visualize it, to communicate it's going to be a hugely important skill in the next decades"

Hal Varian, chief economist Google 2009

Introduction

We are surrounded by data every day in our work and in our play. The data we encounter can be very small, such as a phone number (several bytes), or as large as a corporate database (several petabytes). A decade ago data scientists focused primarily on business intelligence (BI), but now this new science is an integral part of all industries. In the business world algorithms were developed to predict customers who would quit membership (churn) and determine what additional products a consumer might purchase (market basket analysis). Even in less scientific fields, such as sports, they are highly dependent on statistics and data analytics.

In this chapter we will focus primarily on the impact of data science on the healthcare sector. Healthcare data analytics was also discussed by Bill Hersh in chapter 3. Given the nascent nature of this field, there is a need for continuous updates and resources available about the importance of data science and health care data analytics. Being able to work effectively and efficiently with data is especially relevant in this age of healthcare technology. The ease of analysis and modeling of complex and free text data is now key and data science needs individuals with both data science and domain (healthcare) specific expertise. Multiple organizations are supporting data science initiatives and education. Details of data science initiatives are available in Appendix 23.1 of this chapter.

The following are two important definitions:

- Data science *"means the scientific study of the creation, validation and transformation of data to create meaning."* [1] Because data science is relatively new, definitions are still evolving.
- Data analytics is *"the discovery and communication of meaningful patterns in data."* While some would argue for separating data analytics from data mining and knowledge discovery from data (KDD), we will use the terms interchangeably. [2]

Background

The concept of data analytics was outlined by Tukey in 1962 and represented a departure from traditional statistics, which is based on mathematics.[3] The first published use of the term "data science" was in a paper by William Cleveland in 2001, in which he called for expansion of the scope of statistics.[4] Early data scientists worked for some of the most innovative Internet companies, such as Google, Facebook, LinkedIn and Twitter, to assist them with gleaning information from their exponentially growing volumes of data. Data scientists generated new data use cases of data for the consumer and for new business models and products. The advent of the Internet was one of the greatest sources of the data explosion witnessed over the past two decades. For example, in 2014 for every minute there were 4,000,000 Google searches and 2,460,000 shares on Facebook. [5]

In 2013 Cukier and Mayer-Schoenberger aptly named the quantification of data *"datafication."* [6] Almost every Internet activity today can be measured (datafied or quantified) and mined. In addition to the meteoric increase in data volume there is also tremendous variety in the data, such as location data (geographic information system (GIS)), survey data, image data, email data, tweet data and sensor data. The data science field has been greatly facilitated by faster computer processor speed, open-source software designed to process large volumes of data and commodity hardware with more expansive storage. This has led to the Big Data era, which we will cover later in this chapter. For more information regarding the history of data science, Gil Press provided a history of data science from 1962 to 2012. [7]

Clearly, data science is a relatively new field with significant recent popularity, as evidenced by a Google Trends analysis from 2011 to 2016 (Figure 23.1). [8] While there is undoubtedly some hype associated with data science there has been great attention to new courses in data science and the creation of data science centers across the nation. The field is new enough that many details are being worked out among disparate academic departments, such as statistics and computer science.

Figure 23.1 Data science web search popularity 2011-2016 (Google Trends)

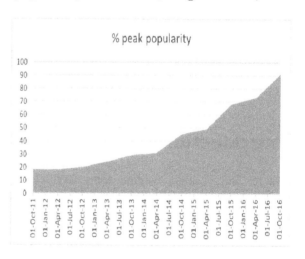

Data scientists are in great demand today and sought after in all industries. Later in this chapter we will cover the educational and career aspects of data science. Data science has evolved rapidly as a field to include many necessary skill sets required by data scientists. While statistics is the cornerstone of data science, there are many more requirements.

The fundamental skill sets and expertise required for data scientists are:

- Mathematics and statistics
- Domain expertise e.g. business and healthcare
- Programming in multiple languages: R, Python, SQL, etc.
- Database and data warehousing
- Predictive modeling and descriptive statistics
- Machine learning and algorithms
- Big data
- Communication and presentation [9]

Given the relative newness of data science it is not surprising that there is some controversy regarding its exact definition and where it fits into the information sciences. The field of statistics has been keenly aware of this new field for several decades, with some statisticians believing they are data scientists, while others believing the field needs to be expanded to include more computer science. Since the 1962 publication by Tukey, there has been a push to expand the education of statisticians to make it more in line with data science requirements. [10]

Biomedical Informatics and Health Informatics have also struggled with the new field of data science. Some authorities feel that well trained informaticists are data scientists, while others maintain that data science training is different and more multi-disciplinary. [11]

The computer science field has evolved with specialized degrees in data science and information science. Multiple universities now offer degrees in data science and we will address this in more detail later in the chapter. Figure 23.2 displays a simple Venn diagram of the field. Data science is located at the center, in the overlap section and is the aggregation of the multiple skill sets.

Figure 23.2 Venn diagram of data science

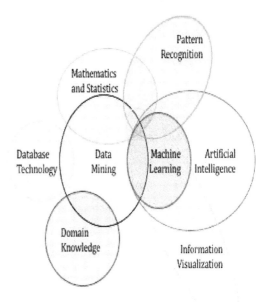

Data Basics

Bits and Bytes: Datum is singular and data is plural. In chapter 2 we covered the definition and importance between the concepts of data, information and knowledge. Data is simply a number without any particular meaning e.g. 10. Information is data with meaning, e.g. a hemoglobin of 10. Knowledge is information that is felt to be true, e.g. a hemoglobin of 14 is normal.

The smallest unit of data is the bit (**b**inary dig**it**) which can be represented as binary choices (0 or 1). A byte consists of eight bits and can provide a potential 256 combinations of data. For example, 0100 0001 represents the capital letter A. Four bytes together would provide more than 4 million possibilities. Because these strings of numbers could be extremely long, the codes can also be displayed as octal (base 8), decimal (base 10) or hexadecimal (base 16). This type of binary coding is important, as computers can rapidly interpret data in this binary format. [12]

With the explosion of data has come the increasing size that challenges us all and is displayed in Table 23.1.[13] What seems like massive datasets today, may seem like medium or small datasets in the future.

Table 23.1 Data sizes

Data Size	Byte Equivalent
Byte (B)	8 bits
Kilobyte (KB)	1,024 bytes
Megabyte (MB)	1,024 Kilobytes
Gigabyte (GB)	1,024 Megabytes
Terabyte (TB)	1,024 Gigabytes
Petabyte (PB)	1,024 Terabytes
Exabyte (EB)	1,048,576 Terabytes
Zettabyte (ZB)	1,073,741,824 Terabytes
Yottabyte (YB)	1,099,511,627,776 Terabytes

Given the many recent "open data" initiatives in existence it is very easy to find and analyze a variety of datasets. In Appendix I of this chapter multiple public-access free datasets are posted.

In order for data to be machine readable it must be in a format the computer recognizes, such as a text file e.g. a comma separated value (.csv) file. CSV and XLS (Excel) files are excellent formats to populate spreadsheets, the most common starting point for data analytics in any industry. While we will not cover spreadsheets in this chapter in detail, it is strongly recommended that students become proficient in applying formulas, math functions, filters, conditional formatting and pivot tables to them.

Statistics Basics

It is well beyond the scope of this chapter to include an in-depth discussion of statistics, so only the salient concepts that relate to data science will be presented.

Data structure: Data can be classified as *structured* (discrete or fits into a defined field in a database or spreadsheet, e.g. text in name field

is a name), *unstructured* (free text) and *semi-structured* (doesn't fit into a database but has a known schema or is tagged, e.g. extensible markup language (XML) file).

Types of data: A classification schema to evaluate types of data would be to categorize it as categorical (nominal or ordinal) or continuous data (interval or ratio). Nominal (also known as categorical) data is *non-numerical* data with no order, such as gender (male or female). Ordinal data is similar, but has order, such as small, medium and large, but there is no numerical measurement difference between the variables. Interval data is *numerical* data with meaningful intervals but no meaningful zero value, such as Celsius temperature. Multiplying interval data does not make sense. Ratio data is numerical and has order, but also has a meaningful zero value, such as height and weight. Multiplying ratio data does make sense.[14] Data can be discrete, such as the number of discharges (no such thing as ½ discharge) or continuous such as weight where the number can be fractional (e.g. 140.5 lbs).

Categorical/nominal data can also be considered qualitative or describing a quality, whereas continuous data tends to be quantitative or numerical data.

Parametric and non-parametric data analysis: Data analysis can also be classified as parametric or non-parametric. Parametric analysis has some form of assumption; is usually associated with a large data sample and the data follows a normal distribution. Non-parametric tests are easier to understand and are usually used for smaller sample sizes. Non-parametric data analytics has minimal assumptions and is used on non-normally distributed data. [15]

Statisticians generally use non-parametric tests (such as a Chi-square test) for non-normally distributed data and parametric tests (such as t-tests) for normally distributed data.

Data scientists will frequently look at measures of central tendency, such as mean, median and

Table 23.2 Data types and central tendency measures

Data	Central tendency measures	Comments
1. Non-parametric		
Nominal (categorical)	Mode	Non-numerical data. Mean not used.
Ordinal	Mode or median	Non-numerical data. Mean not used. Median helpful with outliers
2. Parametric		
Interval	Mean, mode and median	Numerical data
Ratio	Mean, mode and median	Numerical data

standard deviation to see how the data is distributed. The mean is the sum of values divided by the number of values; whereas the median is the middle of the distribution and the mode is the most common value. The range is the difference between the lowest and highest value. The standard deviation is a measure of the dispersion of data from the mean. Note that many statistical methods focus on comparing the means of two different attributes, but the mean of non-parametric data is meaningless (such as the mean of small, medium and large), so other statistical methods must be used. [14] Table 23.2 shows important characteristics of non-parametric and parametric data.

Central Limit Theorem: There are many reasons why data scientists want to know more about their datasets. The central limit theorem posits that if the number of independent random variables is very large, then the distribution should be normal or bell-shaped (figure 23.3).[16] Viewing figure 23.3, you can see that 95% of data that is normally distributed falls within 2 standard deviations and 99.7% falls within 3 standards deviations from the mean. Therefore, values that are greater than 3 standard deviations must be looked at critically to determine if the result is real or an error. An example might be the distribution of the height of all male college students (n=500). The mean might be e.g. 68 inches with a standard deviation of 3 inches. Those students over 77 inches (68 + 9), or under 59 inches (68 - 9) would fall outside the 3 standard deviations. Standard deviation (SD) is calculated by taking the square root of the variance. To calculate

variance, you subtract each data value from the mean, square each result, then add them up, then divide by the number of values -1 (you subtract 1 from the total (n) if you are dealing with a sample, but don't subtract by 1 if you are measuring an entire population). If the variance is e.g. 7, then the standard deviation is the square root of 7 or 2.64. The larger the population, the smaller the variance and standard deviation are likely to be. Also, the curve will be more narrow, so variance and standard deviation measure how much the values vary, or the spread or dispersion of the data. If the standard deviation is very small, there is very little variance and/or the sample (n) is very large. [17]

Figure 23.3 Normal distribution (Wikipedia)

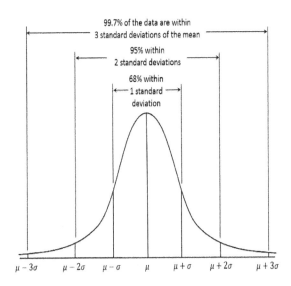

Figure 23.4 shows a distribution skewed to the right (positive) with possible outliers.[18]

Figure 23.4 Abnormal distribution, skewed to the right with outliers

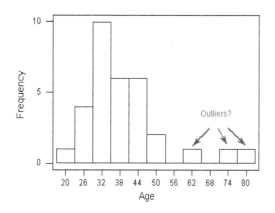

Dependent and independent variables:

Another important concept regarding data analysis is dependent versus independent variables. The outcome (also known as response, label, target or class) variable you are trying to measure is commonly known as the *dependent* variable, such as "developed diabetes" or "didn't develop diabetes." The predictor (explanatory or feature) variables that predict the dependent variable are known as *independent* variables. For example, in your spreadsheet you may have a column labeled outcome where patients either developed diabetes or didn't (dependent variable). Your analysis examines the other columns such as fasting glucose, BMI, HDL cholesterol, family history of diabetes, blood pressure, etc. (independent variables) to see if they predict outcome. In fact, Wilson conducted such a study in 2007 to predict type 2 diabetes in middle-aged adults using logistic regression to examine common risk factors as predictors of outcome. [19]

Descriptive and Inferential Statistics

These are two common approaches by researchers who use data to answer a question.

Descriptive statistics: this approach *describes* the data with several techniques, such as measures of central tendency that includes mean, median and mode. Mean is the most common measure but is distorted by outliers. The median is the midpoint of a distribution of numbers and is less subject to outliers. The mode is the most common value in a dataset. Table 23.2 describes which measures are used with parametric and non-parametric data. Measures of variation are also used, such as range, variance and standard deviation, discussed in a prior section.

Inferential Statistics: analyzes a random sample to infer conclusions about a population. For example, surveys commonly gather information about a sample to predict characteristics about an entire population. Confidence intervals and hypothesis testing are two approaches used with inferential statistics.

- **Confidence Intervals (CIs):** measure the uncertainty of the sample. With any survey or sample one can expect a sampling error or frequently mentioned "margin of error." The most commonly cited confidence level is the 95% confidence level, but other levels such as 90% or 99% could be used. For instance, a 95% confidence interval is a range of values that you can be 95% certain contains the true mean of the population. Using the prior example of 500 male college students with a mean height of 68 inches, a SD of 3 and using 95% confidence intervals, we have 95% confidence that the mean of 68 inches falls between 67.73 - 68.26. The smaller the sample (n) size the wider the confidence intervals. CIs can be determined for numerical and categorical data.

  CIs serve multiple purposes:
  - an easy to understand visual display of the range of values
  - the precision of the estimate
  - comparison of CIs between different studies (e.g. in a meta-analysis)
  - hypothesis testing, with any value outside the 95% CI can help reject the

null hypothesis. Figure 23.5 shows how means and CIs are used to compare three hospitals, using data from the government web site Hospital Compare. CIs are colored gold and indicate they are compatible with national averages. The CIs are narrow, which is optimal, and the actual results can be seen with a "mouse-over" the mean. [20]

o   help "power" a study to determine how many patients a researcher might need to achieve acceptable CIs. [21-22]

- **Hypothesis testing**: measures the strength of evidence to reject or not reject a hypothesis. For example, a researcher has a hypothesis that drug A is better than placebo. The null hypothesis is there is no difference between the mean (μ) of the treatment effect of a drug (stated as $Ho: \mu_1 = \mu_2$) and the alternative hypothesis would be a difference exists (stated as $Ha: \mu_1 \neq \mu_2$). As an illustration, the average or mean systolic blood pressure measured on drug A is μ_1 and the average or mean systolic blood pressure on placebo is μ_2. Commonly, a p-value of .05 is used to determine statistical significance. If the p-value is less than .05 the findings are unlikely to have been caused by chance and we reject the null hypothesis that there is no difference between the two treatments. [14] While p-values are one of the most common statistics to be reported, there are serious limitations and many authorities request confidence intervals and effect size be reported as well.

Effect size (ES): ES measures the *magnitude* of the treatment effect. Unlike significance tests such as the p-value, effect size is independent of sample size. There are a variety of effect size tests. For example, a common test for comparing two independent means is the Cohen d measure which is simply the difference between two means, divided by the pooled standard deviation. A small effect would be d =.2, medium effect d =.5 and large effect d =.8. [23-24] In spite of the known limitations of only publishing significance values, Chavalarias et al. reported that of the 1000 abstracts they reviewed (1990-2015) p- values were reported in only 15.7%, confidence intervals in 2.3% and effect size in 13.9%.[25]

Type 1 and Type 2 errors: a type 1 error is observing a significance (p <.05) when no real difference actually exists (false positive). Remember, p =.05 means there is a 1 in 20 chance for a spurious result. A type II error is when you conclude there is no effect (null hypothesis is not rejected) when one really exists (false negative). The most common cause of a type II error is a sample size that is too small. [14]

Figure 23.5 Hospital Compare: heart failure readmission rates

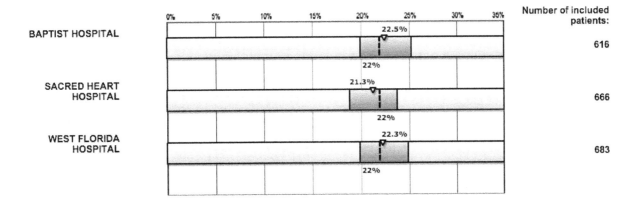

Database systems

Spreadsheets (flat files), such as Microsoft Excel are ubiquitous in many industries, including healthcare and are a logical starting point for exploring and visualizing data. Flat files are two dimensional (rows and columns) and as such are limited, in terms of data analysis compared to a relational database system (RDBS). Most healthcare organizations will use spreadsheets for smaller datasets that require simple analysis and visualization and use RDBSs for larger datasets.

Relational Database structure: Data tables are created that link to other tables to store information in a three-dimensional space (data cube). Table 23.3A shows a simple table with patients as the rows (tuples or instances) and the columns (attributes or features) related to admission and discharge dates. This could link to table 23.3B, a demographic table. Patient_ID in table 23.3B will be the "primary key" which will link to the Patient-ID "foreign key" in table 23.3A.

Cardinality (uniqueness) is an important concept for table construction. A one-to-one relationship means that each row of a table relates to only one row in another table. One-to-many means each row in one table may relate to more than one row in another table. Lastly, many-to-many means that each row in a table may relate to many rows in another table and vice versa. In the example above, Table 23.3A can have the same Patient ID multiple times, while Table 23.3B will have patients uniquely defined.

Traditionally, database systems were designed to adhere to the atomicity, consistency, isolation and durability (ACID) model. Specifically, there needed to be **a**tomicity which is all or none transactions; **c**onsistency or ensuring that if any element of the transaction fails, it all fails; **i**solation means that each transaction is completed separately; **d**urability ensures that each transaction is permanently preserved. However, with the advent of the unstructured NoSQL database models, the ACID model would not pertain. This newer model is known as BASE and will be discussed in the section on Big Data.

Another important concept is "normalization" of database tables, primarily to prevent duplication of data and increase database efficiency. There are 5 defined levels of normalization (also known as normal forms); however, in practice just 3 normal forms are commonly used. The first normal form prevents each table row from containing duplicate data. The second normal form prevents the repetition of data within a table's column. The third normal form requires that every column have a dependency on the table's primary key and independent from the remaining non-key columns in the table. [26]

If databases are being created for a specific purpose they have to be "modeled" or organized so that they are designed to answer important business or clinical questions. In other words, how they are designed will help determine what queries or searches are possible. Databases can be separate (federated) and connected by a computer network or single (central) as in central data repository (CDR). [26-27]

Table 23.3A Simple database table

Patient ID	Last Name	First Name	DOB
69785747	Jones	Roger	01/02/1956
58585758	Smith	Sally	11/15/1940
36637484	Edwards	Edward	03/22/1938

Table 23.3B Connecting demographics table

Patient ID	Hospital ID	Admission Date	Discharge Date
69785747	H445598	2/1/2016	2/12/2016
69785747	H445598	10/3/2014	10/5/2014
58585758	H193240	2/5/2016	2/13/2016
36637484	H148679	2/7/2016	2/14/2016

Database management: Relational data base system (RDBS) can be simple and PC (client) based, such as Microsoft Access and OpenOffice Base. These programs are easy to use but limited in terms of scalability (limit of 2GB data for MS Access and limited user access). They may be appropriate, for example, for a hospital department. Most larger organizations need a commercial database management system (DBMS), such as Oracle or Microsoft SQL Server. Most significant RDBSs are web accessible and use structured query language (SQL) for extracting information, discussed in another section of this chapter.[14] These database systems are classified as online transactional processing (OLTP) systems, where data manipulations are contained in transactions which can be committed on successful processing or rolled back in case of an error.

Clinical data warehouse (CDW): It is likely that a large hospital system may have multiple database systems that need to be integrated so data can be analyzed and reported. This is why organizations frequently establish an enterprise clinical data warehouse. The clinical data warehouse will integrate data from clinical, demographic, financial, insurance, coding and quality-related data systems. They will have, for example, registration, outpatient, inpatient, pharmacy, emergency department and surgical data in one physical location. The data is likely to be both structured and unstructured, to include progress notes, documents and images. While integration of electronic health record data into the warehouse is new in the past 1-2 decades, a majority of the data is unstructured. The goal is to provide a single platform for analytics so that strategic decisions can be made. Data warehouses are used to generate internal and external reports and to perform predictive and prescriptive analytics (see chapter 3 on Healthcare Data Analytics). In addition, the warehouse may interface with clinical decision support systems. Importantly, data warehouses are a great source of data for research.

Many large healthcare systems will also have data marts which are like small warehouses that

focus on a single subject area, such as revenue, and are used by a single department. Figure 23.6 displays a prototypical hospital clinical data warehouse.[28] Enterprise clinical data warehouses can be built from multiple data marts which are connected by conformed dimensions, such as patients, providers, facilities and services provided. A data mart is a collection of data that is of interest to a particular line of business. For a hospital system, the lines of business can include accounting, clinical services, human resources and facility management. The data in a data warehouse is divided into facts and dimensions. Facts are database tables that contain measurements, such as counts and amounts. Dimensions are database tables that contain data that describe the facts, such as name, date of birth and procedure code.

Data warehouses are often associated with online analytical processing (OLAP) that analyzes multidimensional data. OLAP consists of three operations: consolidation (roll-up) or aggregation; drill-down or narrowing a search and slicing and dicing or taking apart the OLAP cube to view from different perspectives. While OLAP helps organize the data, it is not analytical in the usual sense. For that reason, a data mining application must be used. There are programs that combine OLAP with data mining and are known as online analytical mining (OLAM). Most multi-dimensional data models have a star or snowflake schema (architecture). Extract, transform and load (ETL) tools are used to populate and maintain data warehouses. ETL tools allow for the extraction of data from a variety of databases and file types, the transformation of data for cleaning and integration, and then loading the transformed data to the data warehouse.[28] Some of the more popular ETL tools are IBM's DataStage, Microsoft's SQL Server Integration Services and Pentaho's Kettle. These tools provide a graphical user interface so analysts can quickly assemble data processing flows. Furthermore, ETL tools manage the technical details of connecting to databases and opening files.

Figure 23.6 Hospital Clinical Data Warehouse

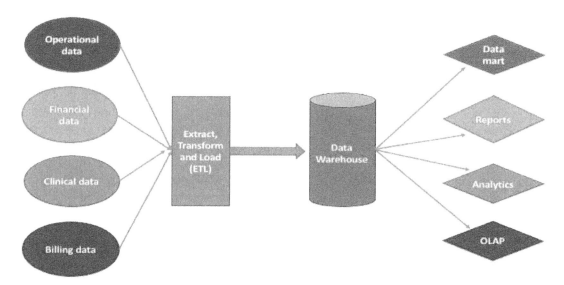

Application Programming Interface (API): Internet-based companies such as Google and Amazon have needed a method to transfer and share data with consumers, developers and researchers. The most common method today is via an application programming interface or API which creates a portal for data retrieval. The most common communication standard today for this is RESTful API. [29] Internal APIs can be created for internal customers and external APIs for everyone else. This technology is also used by web and mobile applications. Healthcare as an industry has been slow in adopting this approach. Open, standardized APIs would make data available to more people on more platforms and overcome the challenge of interoperability. Importantly, developers could innovate and create new operational and analytical tools, not in existence today. [30] There is already an evolving strategy to use the Fast Healthcare Interoperability Resources (FHIR) standard and RESTful APIs to retrieve data from those EHR vendors who have open APIs. The document standards most commonly used are Javascript Object Notation (JSON) and XML. A multidisciplinary group known as the Argonaut Project, that includes some EHR vendors is moving in this direction. They will use new open-source standards

(OAuth 2.0 and Open ID) for secure data exchanges.[31] To further support this new direction, the Federal Government is including access to many data sets via open APIs. For example, there is an API catalog for datasets found on Data.Gov. [32]

Data Analytical Processes

The general steps undertaken with data analysis are displayed in Figure 23.7, but they must be flexible and iterative. The steps in the process are clearly non-linear; they are multiple interdependent cycles. For example, a researcher might explore data in the exploratory data analysis (EDA) phase with a visualization, then perform a simple analysis with statistical modeling which doesn't reveal anything interesting and return to the EDA step to look for more patterns of interest.

Similar to evidence based medicine, data scientists begin with an interesting or important question. The question should be specific, plausible with a potential answer and hopefully solved with available data. If the question is in the form of a hypothesis, the researcher will also have to calculate the sample size of the data to be collected to achieve a meaningful result.

Figure 23.7 Data Analytical Process

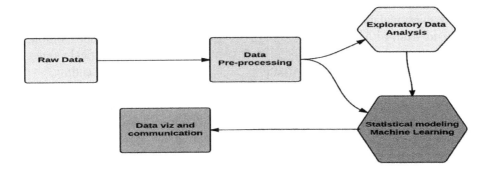

Raw Data

Obviously, the first step is to locate an appropriate data set that hopefully will answer the proposed question. In an ideal world the data would exist in an acceptable and easy to compute format, such as a text file, spreadsheet or relational database. In reality, the data may be unstructured and messy and therefore may need to be transformed into an acceptable format before the pre-processing stage can begin. Raw data (also known as primary data) is therefore unformatted and in its natural state, prior to cleaning or transforming. Raw data from clinical instruments can be stored in the manufacturer's proprietary data format, which typically is not human readable and requires a program to transform the data.

Data Pre-processing

Prior to the actual analysis the data will likely need "wrangling" or "munging":

- Cleaning: correcting and/or removing inaccurate records
- Integration: combining several sources of data into one spreadsheet or database
- Reduction: consolidating categories to reduce the number of attributes [33]

Exploratory Data Analysis (EDA)

EDA builds on the pre-processing phase. The data is examined in more depth, using multiple steps and tools. The main tool used is descriptive statistics to examine the distribution, mean, mode, range, variance and standard deviation (SD), so the appropriate statistical method or machine learning algorithm for eventual analysis can be chosen. Many data scientists claim that roughly 80% of their time is spent cleaning, processing and learning the data, prior to the actual analysis.[9] Visualization of the data is a major component of EDA. Data scientists may examine non-parametric data using pie and bar charts and parametric data using box plots and histograms. Figure 23.8 shows a boxplot of blood pressure. Note the mean, median and standard deviation. The whiskers define the range or minimum to maximum values. The goal is to look for possible skewed data (non-normal distribution) and outliers. Scatter plots are useful to compare two variables for a potential linear relationship. For instance, does a rising BMI seem to correlate with serum glucose? Also, look for missing, incorrect or duplicate data. Will the data set you selected actually answer the proposed question? The

Figure 23.8 Boxplot of blood pressure

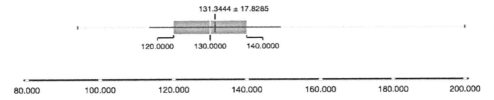

following are additional steps involved with EDA: [34]

- Missing data: most datasets are imperfect, which includes missing data. A strategy must be developed to deal with this, to include imputation and other schemes. [35]
- Standardizing and normalizing data: Data column values that are on differing orders of magnitude or non-normally distributed may need to be "normalized" to prevent one column from biasing or skewing the analytical interpretation. For example, if Column A contains values from 1 to 100 and Column B contains values from 0.0 to 1.0, then these columns must be normalized to put the values in the same range. Software will convert the numerical variables to a scale from 0 to 1. Another scheme is standardization where the numerical mean is zero with a standard deviation of one. This is valuable when your data tends to be normally distributed and you want to use logistic regression or Bayes. Another approach is to convert the data into z-scores by subtracting the observed value from the mean of the data and dividing by the standard deviation. Additionally, z-scores can tell you if an outlier is greater than 3 standard deviations and should be left out of the model.
- Categorization (binning or bagging): If you use a linear (line-like) model, such as linear regression, then the data should be linear. Converting continuous data into bins (e.g. patient age by decade) is also known as

"discretization" and this may help the analysis. Other non-linear techniques, such as decision trees and neural networks are affected less by non-linear data.

- There are times when nominal data must be converted to numerical values, such as 0 and 1; so called dummy coding.
- Variable selection: a data analyst can use a variety of filters in a machine learning software program to see if excluding attributes impacts the results. In fact, part of building a model is to exclude certain variables and see if they impact outcome.
- Segmentation or grouping. Analysts may analyze only a meaningful subgroup and see if that impacts the results. [36]
- Transformation: using a mathematical method (e.g. log transformation) to convert a skewed distribution to a normal distribution or using a Fourier transform to change from the time domain to the frequency domain.[33]
- Table 23.4 displays which visualization and test should be used, based on the type of dependent and independent variables (categorical or numerical).

Analyzing the Data Approaches

Prior to actually discussing analytical methods we will discuss more about analysis in general. There are three general approaches to conduct an analysis: statistical modeling, machine learning and programming languages (see Figure 23.9):

Table 23.4 Recommended test and visualization for dependent and independent variables

		Dependent Variable	
		Categorical	Numerical
Independent Variable	Categorical	Visualize with two-way table and percentages. Test with chi-square	Visualize with side-by-side box plots. Test significance with t-test and ANOVA
	Numerical	Test significance with logistic regression, decision tree or Naïve Bayes	Visualize with scatterplot. Test strength and direction with correlation. Test mathematical relation with linear regression

Figure 23.9 Modeling approach

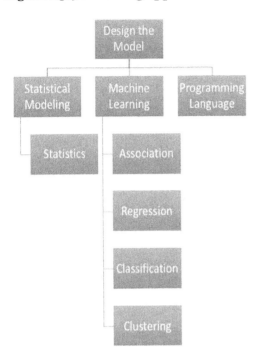

- Statistical modeling is one of the most common approaches and was developed by statisticians. Statistical tools are discussed later in the chapter
- Machine learning approach was developed by computer scientists, but is based on mathematics and statistics. Keep in mind that statistical modeling will also address associations, regression and classification, but not clustering. Modeling with machine learning depends on algorithms which are really mathematical formulas to make predictions from data. For example, there is a decision tree algorithm named classification and regression tree (CART) which uses a mathematical formula to calculate a data column and value that best splits the data so as to separate observations based on the dependent variable. All algorithms learn a mapping from input to output. Models used for prediction are called predictive modeling. Different algorithms are outcome compared. This tends to be different from statistical modeling where one statistical method is usually selected.

Supervised machine learning means you already know the classes of data you will be dealing with. For example, you know that the columns (attributes) will be health and demographic factors and the rows are patients.

Unsupervised machine learning implies you don't know the classification of the data. Techniques such as clustering and associations will discover and report groups or patterns of data. Clustering techniques will be discussed in another section.

Semi-supervised learning means you know some of the classes of data only, the remainder are unknown. Machine learning is based on algorithms to make predictions. Table 23.5 enumerates the various common algorithms used today

- Programming languages. **R language**: The R and Python programming languages are both used in the statistical approach to data analytics, particularly by data scientists. R (statistical computing and graphics) language was developed in 1995 and has gained in popularity ever since. The software is free to download and is supported by a large network and a repository of R functions. This language does include machine learning algorithms. The program tends to be a little slow and there is a substantial learning curve. **Python:** This language was created in 1991 and is considered to be one of the easier languages to master. In general, the R language is used more by data scientists and Python is used more by computer scientists; but considerable overlap exists. Python is a better choice if you need to integrate analytics with a web app or database. There is also a machine learning package (scikit-learn), in addition to the statistical approach. [37]
Structured Query Language (SQL): is a universal language to manipulate databases. Its basic functions to modify tables are **CRUD: C**REATE (new table or database), **R**EAD (select or query), **U**PDATE (edit), and **D**ELETE (remove) information in a

Table 23.5 Machine Learning Algorithms

Algorithm	Examples	Indication	Benefits	Limitations
Linear			Simple, fast and requires little data	Linear model assumptions
Linear regression		Regression	Widely used	For numerical data only
Logistic regression		Classification	Widely used	For binary categorical data
Linear discriminant analysis		Classification		For more than 2 classes of categorical data
Non-linear				Requires more training data. Slow
Decision Trees	CART, CHAID, C 5.0	Classification and regression	Fast, easy to understand	Risk of overfitting
K-nearest neighbor		Classification and regression	There is no learning so entire training set is used. Simple	May have to experiment with K values to optimize. Works best with small # inputs and no missing data. May need to normalize data
Support vector machine	libSVM	Classification and regression	Powerful, handles high-dimensional data. Used for linear and non-linear models	Uses only numeric or dummy coded categorical data; decision boundary can be difficult to interpret
Bayes	Naive Bayes, Gaussian Naïve Bayes	Classification and regression	Fast, often considered a baseline since it uses probabilities	Assumes input is independent. Input is categorical or binary, but numerical for Gaussian
Neural networks	Multilayer Perceptron	Classification and regression	Powerful, robust and generally more accurate	More complex and more difficult to evaluate results
Unsupervised learning				
Distance based clustering	k-means	Pattern discovery where there are no data labels	Simple and easy to implement	# groups must be selected ahead of time, does not handle concave shapes well
Density based clustering	DBScan	Builds clusters based on the density of data points	Finds clusters with concave shapes	A spatial index is required to run efficiently
Association	aPriori, frequent pattern tree	Set of rules defining a pattern where condition A implies condition B	Simplicity	Can be memory and processor intensive
Ensemble	Bootstrapping, Random forest AdaBoost	Classification and regression	Uses several algorithms for more accuracy	Interpretation of the results from multiple models can be challenging

table in a database. Additional important functions are INSERT and TRUNCATE and many more commands are available. Data must be characterized as data types: **char** (size) or fixed length characters with parameter size in parenthesis; **varchar** (size) or variable length characters with parameter size in parenthesis; **number** (size) or number with a max number of digits in parenthesis; **date** or date value; **number** (size, d) or the number value with a maximum number of digits of size and the precision which is the number of digits to the right of the decimal. [38]

Major Types of Analytics

Descriptive Analytics (not the same as descriptive statistics): analyzes data to look for patterns where no target value or class exists or is used. Therefore, this is unsupervised learning and uses <u>three</u> categories of machine learning algorithms:

- *Association rules* are frequently used to associate a purchase of item A with B. They are used to make recommendations and do market basket analysis (MBA). They measure correlation and not causation. Association rules are rated in terms of *support* and *confidence*. Support is defined as the percent of total transactions from a transaction database that a rule satisfies. Confidence measures the degree of certainty of the association. Mining occurs by using the minimal support and confidence levels set by the analyst. *A priori* is one of the most common association rules (algorithms). Multiple rules are generated so you have to narrow them down, based on confidence versus support levels established.[28]
- *Sequence rules* are concerned with the sequence or order of events in a transaction
- *Clustering* identifies patterns or groupings. Hierarchical clustering builds clusters within clusters to assist in comprehending data similarities and reducing the number of representative groupings. Hierarchical clusters can be built from the bottom up

(agglomerative) where each observed data point begins in its own cluster and then clusters are merged as the algorithm climbs up the hierarchy. Another way to construct a hierarchical cluster is from the top down (divisive), with all observed data points in one hierarchy and then having the algorithm recursively split the data into smaller clusters. Non-hierarchical clustering uses techniques such as k-means to calculate distances. The number of clusters (k) has to be established ahead of time. Clustering is generally used with numerical data but algorithms, such as FarthestFirst can handle categorical data. [39] Clustering has been successfully used in the medical field to analyze microarray gene data to look for previously unrecognized relationships. Figure 23.10 shows clustering using k-means algorithm where k = 3 in the machine learning software WEKA. The clusters are based on iris petal length on the y axis and number of samples on the x axis. [40]

Figure 23.10 Clustering with three Identifiable groups

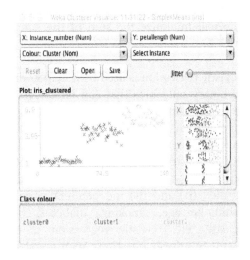

Predictive Analytics (modeling): Creates a model based on a target (dependent) variable, using multiple (independent) variables. Complex machine learning algorithms are not as interpretable or easy to understand. Comments on benefits and limitations of algorithms are included in table 23.5. Modeling approaches include the following:

- **Regression model** is a linear model that uses supervised learning. *Simple linear regression*: this model displays the mathematical relationship between two <u>continuous (numerical)</u> variables. *Multiple linear regression* is for multiple variables. It plots the target or dependent variable (y axis) against an independent variable (x axis). The data is numerical data, such as income or age. The mathematical formula is $y = ax + b$; where **a** represents the slope of the regression line (increase in y divided by increase in x) and **b** is the y intercept or what the y value is when **x** is zero. The goal is to fit the values as closely to the line as possible. For example, if we use the formula $y = 0.425x + 0.785$ and we set x = 2, then y = 1.64. Figure 23.11 shows a simple linear regression using this formula. Note that the line does not fit perfectly. The line is based on "the sum of least squared errors" or squaring the difference between the actual value and the value represented on the line, then summing the errors. For instance, when x = 3 and y = 1.3 the residual or distance from the line is 0.7. The optimal line will minimize the sum of the squared errors. [41]

Logistic regression is used with dichotomous or binary outcomes that can be categorical data or dummy coded as 1 or 0. The independent (predictor or explanatory) variables are interval or ratio data. Outcome can be described with odds ratios. As an illustration, the outcome could be lived (1) or died (0) and the independent variables could be age, smoking, diabetes, heart disease, etc. Logistic regression is very useful to answer questions such as what is the probability of developing diabetes (outcome) for x increase in BMI or age (independent variables)? Note, that although we have included logistic regression under regression model, it is used

in classification discussed in the next section.

Figure 23.11 Simple linear regression (OnlineStatBook)

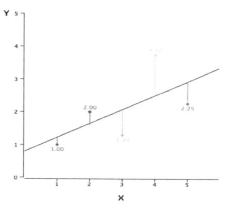

- **Classification model** is a non-linear model that uses supervised learning: unlike the regression model, the target variable is categorical (nominal). Common classification algorithms are as follows:

Decision tree algorithms are now commonly called classification and regression trees (CART). The tree displays root and leaf nodes (outcomes). Examples of decision tree algorithms are C4.5 and J48. Trees can evaluate both nominal data (classification) and continuous data (regression). This is a good way to evaluate attributes, as those with the most information gained will appear first in branching at the root of the tree. Decision trees and regression are the easiest algorithms to explain and present to non-statisticians. These models can be considered a set of IF-THEN rules. Figure 23.12 displays a decision tree looking at contact lens selection. Note, the first branch (root) is based on tear production. IF tear rate is low, THEN no contact lenses. Final decision is no lenses, soft or hard lenses. [40,42] Decision trees can "overfit" the data if they learn the model training data too well; leading to poor predictions on new data presented to the model. Trees often need to be "pruned" with techniques, such as, k-fold validation.

Figure 23.12. Decision tree for contact lens recommendations

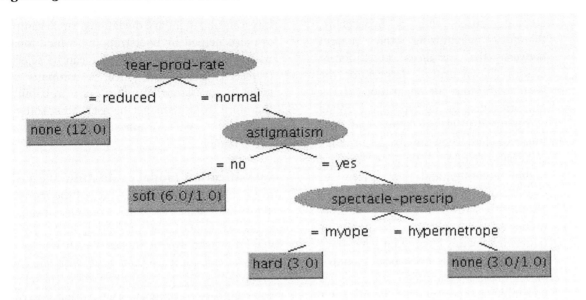

Random forests involve multiple decision tree classifiers (hence the name forest). This is also the reason they are usually organized under ensemble algorithms because they combine several approaches. They work well with multiple predictors and can perform classification and regression. Importantly, a "measure of variable importance" can be studied to determine the relative importance of each variable. Random forests are generally more accurate than standard decision trees, but are more complex.[42]

Neural networks (also known as multilayer perceptrons). While the networks work like the human brain, they are really statistical models, like logistic regression. Results are powerful but more difficult to interpret compared to decision trees. There is also more trial and error associated with this approach due to attempts to optimize the algorithm parameters for the best results. [36]

Naïve Bayes uses Bayes theorem to perform classification and regression, based on probability. Specifically, future or posterior probability is based on prior knowledge or prevalence. It is fast and can evaluate binary, categorical and numerical data. The algorithm is called Naïve because it assumes that each input variable is independent. While this is unrealistic, in reality this approach works quite well. Numerical data can be analyzed using a Gaussian or normal distribution Bayes. Probabilities are calculated based on the normal distribution. [36, 42]

K-nearest neighbor (KNN) is an algorithm used for classification and regression. It evaluates the entire training data set, so there is no learning. Predictions are made for a new data point by looking for the "nearest neighbor". There are several ways to evaluate the distance to a data point, with the most common being the Euclidean distance. KNN does not perform as well on large data sets (multiple input variables); due to the "curse of dimensionality."[36,42]

Support Vector Machines (SVMs) is a technique that separates the attribute space with a hyperplane, therefore maximizing the margin between the instances of different classes. The algorithm determines the best coefficients for separation of the numerical data by the hyperplane. SVMs can be used to classify linear and non-linear models. They can also be used for regression. While a little slow, they are highly accurate and less prone to overfitting. [36,42]

Ensemble methods use multiple models to compensate for inherent deficiencies of any one model. The above mentioned random forest is an example of an ensemble method. More robust ensemble methods will use classification algorithms from the different categories of models listed above.

The above techniques can handle "multi-class" classification where they can evaluate more than two target variables. Classification of two target variables is known as binary classification and is common in health informatics. Examples of binary classification target variables would be (benign/malignant), (hypertensive true/hypertensive false), and (Zika virus yes/Zika virus no). [36,43]

Evaluation of the performance of predictive analytics: It is important to have methods to evaluate the predictive ability of these different analytical methods. Traditionally, with the classification model, the data is split into training (67-80%), used to create and train the model, and test data (20-33%), used to test the model. However, with k-nearest neighbor, neural networks and decision trees all of the data must be used to create the model. In smaller data sets (e.g. less than 1000 observations) cross validation using K-folds (K most commonly is 10) is the best technique. The advantage of this approach is that all of the data

is used for both training and testing. Multiple averages can be calculated as well as confidence intervals and standard deviations. Small samples are also helped by bootstrapping, which means when random sampling a value can be selected (replaced) more than once. A computer can generate a large number of probabilities therefore from a smaller dataset. Most software programs will automatically evaluate performance.

Classification model evaluation: the goal is to establish the overall accuracy of the model or algorithm selected. For example, using a validated dataset for predicting heart disease you know who eventually develops heart disease and you can compare that result with what was predicted using one or more models. This allows you to create a confusion matrix, that describes the performance of the classification model, by displaying the true positives (TP), false positives (FP), true negatives (TN) and false negatives (FN). Table 23.6 displays a confusion matrix based on heart disease data and generated by the machine learning software program WEKA. [40] From that table you can determine the classification accuracy, classification error, sensitivity, specificity and precision. Table 23.7 displays the most important measures and formulas related to calculating classification performance. Figure 23.13 shows the results of a 10-fold validation using three machine learning algorithms. The classification accuracy (CA) was highest with Naïve Bayes. The precision (positive predictive value) was also highest with Naïve Bayes, as was recall (sensitivity). The F-1 measure is the weighted average of precision and recall. Another way to compare these methods is to plot true positives on the y-axis and false positives (1-sensitivity) on the x-axis. A "perfect" result would be 1 where there are only true positives and no false positives. The curve that is created should be in the upper left area and the area under the curve (AUC) should be in the 70-90% range to be a truly significant finding. Figure 23.14 shows an actual ROC curve with Naïve Bayes showing the greatest AUC (.838).

Table 23.6 Confusion matrix for Bayes classifier results

	Predicted No	Predicted Yes	
Actual No	TN = 130	FP = 20	TN + FP = 150
Actual Yes	FN = 23	TP = 97	FN + TP = 120
	TN + FN = 153	FP + TP = 117	Total cases = 270

Table 23.7 Classification performance measures and formulas

Measure	Formula	Result
Accuracy or correct classification rate	$\dfrac{TP + TN}{Total}$	$\dfrac{97 + 130}{270} = .84$
Misclassification or error rate	$\dfrac{FP + FN}{Total}$	$\dfrac{20 + 23}{270} = .15$
Sensitivity, recall, true positive	$\dfrac{TP}{FN + TP}$	$\dfrac{97}{120} = .81$
Specificity, true negative	$\dfrac{TN}{TN + FP}$	$\dfrac{130}{150} = .87$
Precision	$\dfrac{TP}{TP + FP}$	$\dfrac{97}{117} = .83$

Figure 23.13 Ten-fold cross validation results for three machine learning algorithms

Test & Score

Settings

Sampling type: 10-fold Cross validation
Target class: Average over classes

Scores

Method	AUC	CA	F1	Precision	Recall
Naive Bayes	0.838	0.841	0.819	0.829	0.808
kNN	0.652	0.659	0.603	0.625	0.583
Classification Tree	0.749	0.752	0.722	0.719	0.725

Figure 23.14 Receiver operator characteristic (ROC) curve

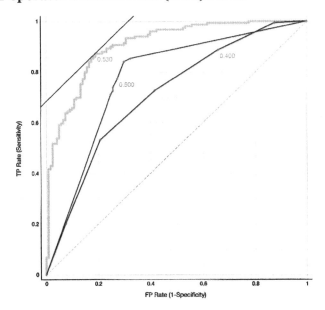

Another choice is a lift curve which gives similar results. Figures 23.13 and 23.14 were created using the machine learning program Orange®, discussed in another section.[44]

Regression model evaluation: The performance is usually measured using R (correlation coefficient) and R^2 (coefficient of determination). R measures the strength and direction of a linear relationship, such that it can be in the range from -1 to +1. R^2 determines how close the data fits the regression line and can vary from 0-100%. 100% indicates that the model explains all the variability of the data around its mean. Additionally, the root mean square error (RMSE) is used as a measure of the fit; the lower the RMSE, the better. [36]

Do the results make sense, it terms of magnitude, direction and uncertainty? The results of the above tests may cause the data scientist to tweak the algorithm to see if the optimal predictors were used. Similarly, running the model with fewer parameters to see if that affects the outcome is reasonable. Additionally, it is reasonable to re-run the model without outliers to see if that impacts outcomes.

If a confusion matrix is calculated, a decision must be made regarding the ideal sensitivity and/or specificity desired. As an illustration, if the goal of the algorithm is to not miss a diagnosis of a serious condition, then a high sensitivity is important. If the goal is to be sure a negative test means no disease, then high specificity is important. [45]

Putting it all together

There are multiple scenarios where predictive modeling in the medical field would be useful. Predicting patients at risk of morbidity (e.g. early sepsis detection), mortality, readmission, high cost and poor compliance are just few of the reasons prediction is important. While retrospective (after the fact) prediction is common, ultimately the goal is to predict serious situations real-time. In the latter scenario predictive analytics = clinical decision support; which will be discussed in chapter 24.

It is important to understand the concept of modeling. An analogy might be baking a cake is the model and the recipes might be the machine learning algorithms or a statistical approach; with multiple ways to bake a cake. After selecting an appropriate data set, the researcher develops a question or hypothesis. He or she first explores the data set to clean it up and then explore it to look at whether there is missing data, whether the attributes are normally distributed and whether the scales of the attributes are similar or dissimilar. If the outcome variable is nominal, then a classification model is indicated. One option would be a statistical approach. In the case of machine learning (ML) there are a multitude of classification algorithms available to build a model. In addition, more than just raw data can be analyzed. The same algorithms can be used for normalized, standardized, and imputed data, and with data sets where the attributes have been reduced to see if that alters results. The overall results can be compared with t-tests to see if there is any statistical difference between the results. ML results can be analyzed for sensitivity and specificity as well. Once a model has been selected, the algorithms can be further tweaked for the best performance, as they all have customizable options. Finally, the model is saved and used for analysis of similar new data.[46] Case study A illustrates many of these points.

Natural Language Processing and Text Mining

The last area of data analytics we would like to mention is text mining using natural language processing (NLP). Natural language processing technology, refers to the ability of artificial intelligence to extract information from natural human language. NLP applications are based on the linguistic-artificial intelligence intersection of computer processing that was developed to

Case Study A: Predictive Modeling with a diabetes data set

A data set hosted on Vanderbilt's Department of Biostatistics web site was used that contains lab and demographic factors on 403 African-American adults from 2 rural counties in Virginia. (See data sets in the data science resource appendix). Thirteen patients were excluded due to missing labs. Sixty patients were diabetic out of the total of 390 or 15%. Diabetes was diagnosed based on a lab test (A1c) of 7 or greater. The outcome (dependent variable) was diabetes or no diabetes. The other 14 attributes (independent variables) were glucose, total cholesterol, HDL cholesterol, Cholesterol/HDL ratio, age, gender, height, weight, BMI (calculated), systolic blood pressure, diastolic blood pressure, waist size, hip size and waist/hip ratio.[47]

Research question: what features (attributes) in the data set best predict the presence or absence of type 2 diabetes in this cohort?

Statistical predictive modeling: IBM Watson Analytics (IBMWA) (discussed in later section) was used. Data was first evaluated with IBMWA's *Explore* option and it was noted that not all variables were normally distributed. Data was then evaluated with IBMWA's *Predict* option. The single strongest predictor (90%) was glucose, followed by age and systolic blood pressure (85%). Logistic regression was used and all calculations had significant p values, but glucose also had a large effect size. When a combination of factors was analyzed, glucose and weight, glucose and waist/hip and glucose and age all had good predictive values but they were not statistically significant. [48]

Machine Learning predictive modeling: WEKA machine learning software (discussed in later section) was used. *Exploration*: There are 390 instances (rows or patients) and there was no missing data. The independent variables were first visualized to determine if they were normally distributed and about half were. The scales for the attributes were different, thus normalization of the data should be considered. The outcome variable was not balanced (many more non-diabetics than diabetics) so another adjustment (weighting) may be in order. All attributes were visualized with a scatter plot and the outcome was not clearly separable, another reason the data should be evaluated with different views. *Prediction*: given that the outcome variable was nominal, several common classification algorithms were selected: Random Forest, Logistic Regression, Naïve Bayes and K-Nearest Neighbor. The 4 algorithms were used to analyze 1. Raw data 2. Normalized data 3. Standardized data 4. Top 4 ranked attributes data. The results showed that normalizing and standardizing data did not result in better performance. Reducing the attributes from 14 to 4 had little effect on performance of the model. Logistic Regression was the best performer but not statistically different from Bayes or Random Forest. It successfully classified non-diabetes 97% of the time, and diabetes 91% of the time with a 4% standard deviation. Using just glucose as the primary predictor had high sensitivity and specificity.[40]

Conclusion: The statistical and machine learning approaches resulted in similar results. We used logistic regression in both, but machine learning provided more choices. Glucose was the strongest predictor of diabetes. IBMWA provided preliminary predictions faster than WEKA, but the latter provided more information about accuracy (such as precision, recall, ROC, etc.). The machine learning model selected was logistic regression and we know that in the future we could use glucose only in the model for prediction.

identify semantics and spoken prose.[49] Health care data, especially electronic health record (EHR) data, is largely unstructured, so perfectly ripe for text mining, ready to reap health care information from "troves of unstructured data."[50] Since clinical notes often contain bullet points, lack complete sentences, contain ambiguous words and acronyms, health care data does pose a challenge for natural language processing. Natural language processing and text mining have the potential to address clinical outcomes and billing, as well as improve interoperability during transitions between electronic health records.[51-52] A systematic review reported in 2016 concluded that NLP mining of EHR text resulted in improved case detection rates when combined with ICD-9/-10 coding. [53] The demand for text mining is increasing, and global market for NLP is predicted to increase from $1.1 to $2.7 billion, driven by the quick evolution in NLP technology, reduced cost and increases in healthcare unstructured data. [54]

Text mining has become more achievable with open-source software now readably available. An example would be the R language (discussed in a prior section). For example, there are packages ready-to-install into R that aid in text mining, such as the *tm* package. [55] Other open source software that can be used for text mining are Apache, Stanford's core, and GATE. [56] There is also an open source consortium to advance efforts and collaboration in natural language processing in healthcare. [57] More resources for natural language processing and text mining are available in Appendix I of this textbook. Case study B demonstrates how using NLP on EHR data can help identify heart failure patients

---

Case Study B: Automated ID and Predictive Risk for Heart Failure Patients

Heart failure (HF) admission and readmission are quite common and costly for patients and healthcare systems. Many institutions are using analytics to try to identify patients at increased risk of morbidity, mortality and readmission. Ideally, patients should be identified while still in the hospital and aggressive treatment and education administered.

Research question: can predictive analytics using natural language processing identify patients better than relying on only manual chart reviews?

Predictive analytical approach: Intermountain Healthcare mined free text dictated reports residing in their EHR daily (uploaded to the data warehouse), using natural language processing (NLP). The goal was to identify and risk stratify current patients hospitalized for heart failure (HF) and to determine if adding NLP would improve on manual chart reviews. The statistical model calculated their 30 day all cause readmission risk and 30 mortality risk. The prediction model for identification of HF patients consisted of these predictors: diuretic use, b-type natriuretic peptide level > 200pg/ml, ejection fraction less or equal to 40 in the previous year and ever eligible for CMS or Joint Commission HF core measures.

Results: The addition of NLP to coding increased the identification score from 82.6% to 95.3% and specificity from 82.7% to 97.5%. This resulted in a HF Risk Report reviewed daily in targeted discharge planning sessions and a reduced amount of time for clinicians to review HF patient's risk, compared to the manual method. It should be noted that Intermountain Healthcare tested this strategy at only one hospital and while they reported a significant reduction in 30-day mortality rates and an increase in the percent of patients discharged to home, they did not see a reduction in readmission rates.

Conclusions: NLP improves HF identification rates using EHR free text. Leveraging technology and targeted discharge planning can result in improvement in HF morbidity and mortality. [59]

Visualization and Communication

Data scientists will likely need to visualize data more than once during the complete data analytical process. Data visualization assists in the understanding and analysis of complex data by placing the data in a context that is easier to perceive. The visualization must present the data accurately and concisely, and be free from distractions. [58] They may first visualize data during the EDA phase to look for patterns of interest. After completion of the formal analysis it is quite likely the findings will need to be communicated to others, particularly those working in the C-suite (CEO, COO, etc.). "A picture is worth a thousand words" also applies to presenting complicated analytics. A dashboard may need to be created that displays a variety of charts, depending on the nature of the data. Graphs and tables can be created with any spreadsheet software but may not offer as much functionality as data visualization software discussed in another section. Microsoft Excel 2016 offers the following graph tools:

- Column (vertical) charts of several varieties to include 3-D
- Bar (horizontal) charts of several varieties to include 3-D
- Pie charts
- Line charts of several varieties to include 3-D
- Scatter charts and Bubble charts
- Area charts
- Sparkline charts which are tiny charts within one cell that summarize data in a line or column [60]

As stated previously, bar graphs are excellent choices for categorical (nominal) data, as are pie charts, if the user is attempting to show the breakdown of components contributing to the whole. Line graphs (run charts) are useful to display numerical data (y axis) over time (x axis). Histograms display continuous numerical data, frequently in bins, such as age groups or decades. Unlike bar graphs the bars are touching. Scatter plots or diagrams are helpful to show the relationship between two numerical variables. Data displayed in this manner may help to show a positive or negative relationship between variables. Summary tables are good choices for both categorical and numerical values. For more specific guidance to presenting tables and graphs, readers are referred to the textbook by Horton. [61]

A newer presentation strategy is Infographics or a mashup of graphs, charts and text to make a strong point. Displays can be web-hosted so the data is interactive and available to anyone with Internet access. A variety of visualization software can create infographics as discussed in another section. Piktochart is a company that offers free and paid accounts to create infographics with about 400 templates. [62]

Big Data

Clearly, in the data science and analytics fields no term has received as much attention and hype as "Big Data". Even the definitions tend to be controversial. The following are two very different definitions of big data:

- Data so large it can't be analyzed or stored on one computational unit [28]
- Five Vs: the reality is that the definition started with three Vs but has increased to five:
  o Volume: massive amounts of data are being generated each minute
  o Velocity: data is being generated so rapidly that it needs to be analyzed without placing it in a database
  o Variety: roughly 80% of data in existence is unstructured so it won't fit into a database or spreadsheet. There is tremendous variety, in terms of the data that could potentially be analyzed. However, to do this requires new training and tools.
  o Veracity: current data can be "messy" with missing data and other challenges. Because of the very significant volume of data, missing data may be less important than in the past.

Figure 23.15 Big Data platform

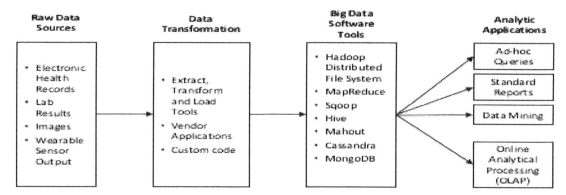

o   Value: data scientists now have the capability to turn large volumes of unstructured data into something meaningful. Without value, data scientists will drown in data, not information or knowledge.[63]

Big data is different from "little data" for several reasons. With big data, domain specific knowledge is of major importance due to the unstructured nature of big data. Collaboration between domain specific researchers (health care), computer scientists, statisticians, and data scientists ensures that knowledge is accurately extracted from the copious amounts of unstructured data. Data science is focused on interpretation of the data and dissemination of the results.

Google developed MapReduce in 2004 which is a system to deal with huge datasets. Later the Hadoop Distributed File System (HDFS) was developed which is the open-source version of MapReduce plus the Google File System (GFS). MapReduce has two functions, mapping and reducing which works by distributing the work among many computers (nodes or clusters). Google routinely analyzes data with 1000 computers, as an example. Although Hadoop is "open source" and free to use, Cloudera provides management tools and offers support contracts. By default, the HDFS creates three copies of the data as it is ingested into the file system, which ensures that the data is always backed up in case of hardware failure.   The most common approach to analyzing big data is to actually reduce the volume of data down to something more manageable and to distribute the workload and data storage across multiple computers by taking advantage of parallel processing. MapReduce runs on the HDFS and, as its name implies, MapReduce first creates a data map of key-value pairs (tuples) and then reduces the data by combining the key-value pairs into a smaller set of tuples. [64] Apache Hive is the data warehousing system built on Hadoop. [65] Apache Hive allows structured query language statements to be executed to retrieve data from the HDFS, allowing a HDFS novice the ability to query and analyze data without the need to fully understand the complexities of the file system. In this regard, Apache Hive is similar to relational database query engines. Apache Mahout is a machine learning library that can also be used with Hadoop. (DDS).[66] Figure 23.15 shows a typical big data platform (adapted from [67]).

In a clinical environment, medical professionals collect patient data such as heart rate, blood pressure and temperature, and also enter codes for a diagnosis. Also, most test results are reported in numerical quantities. These values are easily stored in columns within a database table row which makes for straightforward data analysis and pattern recognition. Where big data processing enters the picture is when there is a need to analyze any notes that are manually entered.   These notes are considered text documents that contain unstructured data. While medical terminology attempts to be standardized,   there   are   still   synonyms, shorthand, dialect and language differences that complicate pattern recognition in text documents. It is important that pattern recognition and text mining software can handle the semantics within the text to attain the meaning the author is attempting to convey.

As an alternative to the Hadoop Distributed File System and traditional relational databases, NoSQL databases can also store and retrieve large volumes of unstructured data. NoSQL is a general term for a variety of database technologies that were created due to the exponential rise in the volume of generated data, the increase in data access frequency, and the performance and processing needs this surge in data volume requires. Relational databases, with their roots in 1970's technology, were not designed to accommodate the demands of scaling up nor the agility that modern applications require. Furthermore, traditional relational databases were not designed to leverage the inexpensive storage and processing hardware available today. NoSQL database systems are different because their non-relational and distributed architecture allows for the quick, impromptu arrangement and investigation of high-volume, dissimilar data types.

The term NoSQL is actually a misnomer since many of the NoSQL databases have a structured query language to manipulate the data. For example, Apache Cassandra has its own query language named Cassandra Query Language (CQL) with commands that are similar to ANSI SQL. Sometimes NoSQL databases are called cloud databases because of their prevalence on cloud service providers, non-relational databases, or Big Data databases. NoSQL databases are specialized in terms of the types of data they were designed to accommodate.[68] The aforementioned Apache Cassandra [69] is designed to store key-value pairs while MongoDB [70] is well suited to store large volumes of unstructured text documents.

While traditional relational databases are governed by the ACID principles (atomicity, consistency, isolation and durability), NoSQL databases use the principles of BASE. In this play on words from chemistry and pH levels, BASE is an acronym for Basically Available, Soft state, Eventual consistency. Basically Available takes advantage of the distributed system architecture so that data is always available even though parts of the system might be offline. Soft state means that the data does not need to be immediately consistent and inconsistencies can be tolerated for specified time periods. Eventually consistent means that after an allotted amount of time, the database will return to a consistent state.[71]

Another application of big data processing in a health care environment is with pattern recognition in digitized images such as x-rays, ultrasounds, CT scans and MRIs. As the resolution and image sizes continue to increase, it is imperative that hardware and processing techniques maintain their ability to efficiently analyze this image data. Image data volume can be reduced through the use of wavelet transforms and other compression algorithms to decrease pattern recognition algorithm processing times. Healthcare big data analytics faces many challenges. Only large healthcare organizations are likely to be able to afford a big data center and have the expertise to run it. Even when using leased hardware from a cloud service provider to reduce hardware costs, the personnel costs can for big data analytics expertise can be prohibitive. Manipulating large volumes of unstructured data will be difficult, except for those with significant training and experience. Integrating genomic information will eventually become routine, but currently, there are multiple questions about how to use the data. There is inadequate physician training in genetics and a shortage of geneticists to counsel patients. (see chapter 20 on Bioinformatics). Social media data is interesting but "messy" with many missing values. Sensor data, particularly from activity monitors is increasingly common, but how can the data be aggregated and which payer will reimburse for this type of monitoring? Unique challenges are also associated with image, geospatial and temporal data.

According to Gina Neff, the biggest challenge "is social, not technical". She points out that the data needs of physicians, patients, administrators, payers and researchers are very different. Most clinicians are likely to tell you that they have lots of data but very little information and very little time and expertise to analyze these data. She also questions how the analytical results will be incorporated into the daily workflow of busy clinicians. Patients have similar issues and concerns, as evidenced by the number who download mobile apps, but do not use them long term. Patient privacy and security are also an extremely important issue with large datasets. [72]

Analytical Software For Healthcare Workers

Statistical software programs

- Microsoft Excel with Data Analysis ToolPak. The Toolpak is a free add-in for Excel users. Most of the commonly used statistical tests are offered. There is a steep learning curve, as users must select the optimal test with no guidance or wizards. [73] A target attribute must be compared with one independent variable at a time, which is labor intensive. There is also a data mining add-in for Excel that connects to an existing Microsoft SQL Server Analysis Services, discussed in the data mining section.
- SQL Server Analysis Services has multiple tools for statistical modeling and machine learning that include both supervised and unsupervised learning: clustering, factor analysis, Bayesian and neural networks, time series analysis, association analysis, recommendations, and shopping basket analysis scoring binary outcomes and linear regression. Wizards are available to expedite the process and assist in configuration. [74]
- Statistical Package for the Social Sciences (SPSS). There are many stat packages available, but we will only concentrate on one. SPSS is a family of analytical packages: *SPSS Statistics*: covers all of the basic

statistical approaches to include temporal causal modeling, geospatial analytics and R programming integration. *SPSS Modeler*: is a predictive analytics platform that includes machine learning algorithms. *SPSS Analytic Server*: accepts data from Hadoop and the Modeler to analyze big data. *SPSS Social Media Analytics:* analyzes Twitter feeds [75]

- IBM Watson Analytics (IBMWA) was released in 2015 for data analysis by any industry. It is unique, in that it automates the exploration, prediction and visualization of data. The analytical approach is based on advanced statistics (SPSS) and natural language processing (NLP), and not machine learning. NLP tools automatically generate 10 observations about the uploaded dataset with the ability to use NLP to ask additional questions. Data quality is scored and associations are noted. In the background under "statistical details" a user can see the test used, its indication, the significance value (p-value) and effect size. Visualizations are automatically created and can be later used to create dashboards and infographics. An academic program is available for teaching so that an instructor and his/her students can use the program without charge. [45,76] Figure 23.16 shows a prediction result looking for predictors of obesity, based on County Health Ranking Data, for the state of Florida.[77]

Machine Learning/Data Mining
- SQL Server Analysis Services (see prior section)
- WEKA: Developed by the University of Waikato in New Zealand, this software is associated with its own textbook and online course. It differs from the other programs by not having a graphical user interface that requires users to move operators (widgets) and connectors around to run an analysis. It performs classification, association and clustering. Data visualization is available but is primitive, compared to newer programs.[40]
- KNIME: is a suite of open-source analytical tools. The program supports a large variety of data types and connectors (algorithms

Figure 23.16 Predict option in IBM Watson Analytics

and steps). Integrates with programming languages, such as R and Python. More than 1000 modules are available, which is both a strength and weakness, because there is a lot to learn before starting. Software is available for all operating systems. The user must select the correct operators and sequence for this to work properly. A Quickstart Guide is available. [78]

- RapidMiner. They offer a suite of paid and free machine learning software, available for all operating systems. The free version is RapidMiner Studio, but they also offer a server based solution intended to be integrated into an enterprise platform. Excellent tutorials are available. This is a very comprehensive system, but you must know which operators to drag and drop, and how they are connected, in order to get results. [79]

- Orange: Developed by the University of Ljubljana in Slovenia, this program has good tutorials and instructions. Major categories are Data, Classification, Regression, Clustering, and Visualization. Software is available for Windows, Mac and Linux OSs. Probably the most intuitive choice of the free open-source machine learning software. Right clicking operators tells the user what they need to know and gives examples. Figure 23.17 shows a screen shot of a machine learning exercise using three common algorithms. [44]

Figure 23.17 Machine learning using Naïve Bayes, nearest neighbor and classification tree

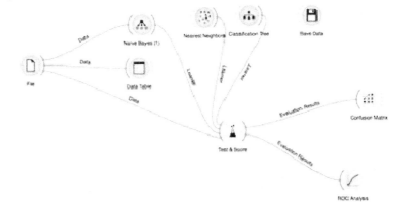

Data Visualization. Only two data visualization software programs will be presented. Both suggest that they perform analytics but in reality they don't perform descriptive or predictive analytics, just visualization.

- Tableau: The program offers a suite of web-based and client-based software. There is a free version known as Tableau Public, as well as paid versions. The Public version is non-web based and available for Windows and Mac computers. Input can be Microsoft Access files, Excel or .csv files. Software is very intuitive, with drag and drop functionality. The visualization choices are similar to Excel, but more elegant and with more options. There is also a free reader for Tableau program output. Tableau offers white papers to improve visualization skills. [80-81]

- Qlik SenseDesktop: This is the free version of the software, which is only available for Windows operating systems. Like Tableau it has an easy drag and drop functionality and multiple chart formats to choose from. They offer developers an open API option. [82]

Data Science Education

DataScience Community is a web site that displays colleges and universities offering a data science-related degree, such as data mining, data analytics and data science in the US and overseas. An early 2016 query revealed 379 institutions in the US offering such programs. The breakdown is quite varied and is as follows:

- Certificate (82); Bachelor (24); Masters (259); Doctorate (14)
- 37% of courses are offered online
- 101 programs were related to business schools, 40 related to mathematics and statistics departments, 39 related to computer science departments and 9 related to new data science departments. The remainder were from a wide variety of college and university departments. [83]

KDNuggets is an excellent resource for US and international online courses in analytics, big data, data mining and data science. They also have sections on certification and certificates in analytics and data science. [37] Certifications:

- Certified analytics professional (CAP): CAP has offered certification since 2013. The CAP® credential targets professionals with 5 years of professional analytics experience and a bachelor's degree, or with 3 years of experience and a master's degree. [84]
- Certified Health Data Analyst (CHDA): hosted by AHIMA, the eligibility requirements include: Registered Healthcare Information Technician (RHIT) credential and a minimum of three years of healthcare data experience; Baccalaureate degree and a minimum of three years of healthcare data experience; Registered Healthcare Information Administrator (RHIA) credential; Master's in Health Information Management (HIM) or Health Informatics from an accredited school or Master's or higher degree and one year of healthcare data experience. [85]

In addition, Coursera in 2016 lists 141 low-cost courses related to data science. There are a variety of short courses that would allow someone to see if they have interest or potential in this new area. [86]

The previous paragraphs suggest that there are numerous conventional and non-conventional data science programs available and reflects the overall attention this field has received recently. At the current time most students will likely opt to get a Master's degree, but have to find a way to gain domain experience, such as internships and employment.

In addition to a plethora of new data science courses, there are also an abundance of new Data Science Centers appearing around the US. An informal search found thirteen major centers with many being multi-disciplinary. This is an encouraging trend because in years past it would have likely that they would have belonged to either computer science or math and statistics departments. Universities are realizing that data science/data analytics affects all industries and therefore affects multiple schools and departments. Perhaps the most striking data science center plan was announced by the University of Michigan in 2015. They anticipate spending $100 million over the next five years to create their Data Science Center for students and faculty and hire 35 new faculty members. [87]

Data Science Careers

Multiple jobs are available in the data science and analytics field. An early 2016 query on Indeed.com for "data scientist" revealed 2,457 jobs and a query for "data analytics" returned 13,196 jobs with considerable overlap between the two specialties. Both were associated with salaries between $75K to over $125K. [88] A 2015 O'Reilly Media Data Science Salary Survey surveyed 600 individuals in the data science field. Two thirds of respondents were from the US and only 25% were actually data scientists, the remainder were employed in a variety of data-related positions. The survey also revealed that 80% of employees in the field were male; having a PhD added $7500 on average and management roles also added to salary estimates. [89] It is clear that most data science-related employees had a Master's level degree, but experience, curiosity and problem solving skills were equally important.[90]

Large healthcare organizations such as Kaiser-Permanente have a large data analytics workforce. According to a Kaiser-Permanente director of information the most important skills are programming, communication and critical thinking. The general data employment categories are:
* Data analysts: technical group who deal with the architecture
* Report analysts: design reports and dashboards related to key performance indicators (KPIs)
* Business systems analysts: work with the operational aspects of business

* Informatics analysts: have strong statistics background and who help others with data analysis [91]

KD Nuggets posts a non-academic and academic data science jobs board. Companies such as Verizon, Aetna, Geisinger Health System, Schwab, Microsoft and Sears are just a sampling of companies seeking data science and analytics expertise.[37]

Data Science Resources

See Appendix 23.1 at end of chapter

Data Science Challenges

The primary challenge facing the new field of data science is training enough data scientists. According to a 2011 McKinsey Global report there will be a shortage of 140,000 - 190,00 data scientists with significant analytical skills for all industries, by 2018.[92] It is unclear whether a single Master's level training program is adequate. Clearly, some individuals will need additional training and many more will need experience in a domain, such as healthcare or business. Actual exposure to complicated data situations and big data will be critical for success. Due to the shortage of data scientists, it is likely that many organizations will have to rely on a team with multiple individual skill sets, rather than a single individual with all of the required skills. It is possible with better automation and simplification of data analytics; shorter training may be acceptable.

It is also challenging to educate people involved with data to use both statistical and machine learning tools. Both are important and both should be used to expand the analytical tool set. Classic statistical tools such as linear regression may be too rigid for evaluating biological data that is more complex with more unknowns. This is often better analyzed with machine learning algorithms. Newer data challenges such as image or speech recognition have demanded newer methods, such as neural networks.[93]

Many have argued that medical data analytics is unique and not similar to data from the physical sciences. They cite that the data is more heterogeneous and complex (free text physician, lab, pathology and xray reports); associated with more social, ethical (privacy and security) and legal issues; subject to physician interpretation of results and more difficult to model than other non-healthcare data. In spite of the information being electronic, it can still be inaccurate and incomplete. [94]

Another challenge facing the data science field is hype. While there is no doubt that we need to train more people in data analytics, it is unclear how often big data analytics is actually required. Challenges with big data were covered in that section. Finding correlations, using statistics or machine learning is not difficult, but correlation doesn't prove causation. For example, fire engines don't cause fires, just because they seem to be omnipresent at fires.

Future Trends

It is likely that automation of analytics, such as IBM Watson Analytics will become more common place and will allow for more healthcare workers to be involved in data science and analytics. A 2013 survey by KDNuggets sought to find what users thought of the likelihood of automated predictive analytics: 5.1% thought it already existed, while 46% thought it would be 1-10 years off, 30% that it would take another 10-50 years and 19% thought it would never happen. [37]
More and more data analytical platforms are becoming available. In addition to IBM, Amazon and Microsoft, Google began offering the Google Cloud Platform in early 2016. This comprehensive platform offers a myriad of tools (26) that fall into the following categories: compute, storage, networking, big data, machine learning, operations and tools. [95]

In David Donoho's sentinel paper "50 years of DS" he posited the following future trends:

- Open science: more sharing of data and code among researchers to promote reproducibility
- Science as data: verifiable results, available to all researchers
- Data analysis tested empirically: rather than using mathematical models under ideal conditions, analyses will be judged with empirical methods. [9]

We will continue to see refinement in many areas of data science. It is likely that future data scientists will successfully blend statistics with machine learning and not treat them as completely separate approaches. Not only will the definition of the field likely see changes, the training programs will likely evolve as well. Microsoft will offer a free six-week course in data science beginning in July 2016 as just another example of industry promoting this new field. [96]

Data science will continue to be promoted by the federal government because the field is critical to the future direction of healthcare. *Precision medicine* is "identifying which approaches will be effective for which patients based on genetic, environmental and lifestyle factors" which will require the integration and analysis of extremely large and complex data sets. [97] This will be one approach to achieve a *learning health system* or "an ecosystem where all stakeholders can securely, effectively and efficiently contribute, share and analyze data." [98]

Given the focus on big data, one can expect improvement in processing, analyzing and storing huge datasets. More data centers are likely to appear, with specialization by industry.

Moore's Law will continue to drive miniaturization and improvement in speed, memory and storage. [99]

Key Points

- Data science is a new information science field with significant overlap with other fields

- A data scientist must have multiple skills, in addition to mathematics and statistics

- There is a shortage of data scientists worldwide

- Healthcare needs data science and analytics, like all other industries

- Big data brings new promises and new challenges

- We need new tools to automate the process of descriptive, visual and predictive analytics

Conclusion

Data science is an exciting new field that is broad and incorporates multiple disciplines. The field mandates multiple skill sets that might not be achieved with only one Master's degree. Given the current shortage of data scientists, many healthcare organizations will have to rely on a team for the required expertise in multiple areas. In spite of numerous training programs in data science, it will take a considerable amount of time for there to be adequate data scientists in the pipeline. In addition, most graduates will need experience in the field (domain) to develop expertise. Healthcare organizations need to look at potential internships in the CIO's arena to gain the necessary experience to intelligently analyze healthcare data.

Colleges and Universities are already "ramping up" in the data science field to educate more workers in multiple industries about data analytics, statistics and machine learning. Healthcare workers who are unable to enroll in local data science courses have a plethora of low-cost online national courses available.

It seems likely that we will see more automation of analytics in the future. Data visualization is perhaps the easiest area to automate, so that is why we are seeing vendors expand their influence in this area. It is likely we will see vendors produce new products for predictive analytics, in addition to IBM Watson Analytics. Importantly, new products will need the means to evaluate and report their results. In healthcare, performance is most commonly reported out as sensitivity, specificity, predictive values, likelihood ratios (LRs) and area under the curve (AUC). It is just a matter of time before we see this type of robust analytical program embedded into healthcare data sets.

Appendix 23.1 Data Science Resources

General Datasets	Details
• University of California, Irvine Repository: https://archive.ics.uci.edu/ml/datasets.html	Site includes 325 validated datasets covering many domains, different sizes and data types and different analytical methods. These data sets are commonly used for machine learning exercises.
• KDNuggets: www.kdnuggets.com	Site includes 71 data sets available for free download, from various industries.
• The Datahub: https://datahub.io/dataset	Managed by the Open Knowledge Foundation, this site hosts more than 10,000 datasets from most industries.
• Kaggle: www.kaggle.com	Provides free, interesting datasets for various user interests and analysis.
Healthcare Datasets	**Details**
• Healthcare Cost and Utilization Project (HCUP): http://www.hcup-us.ahrq.gov/	Includes U.S. longitudinal hospital care data with databases, software and online tools
• Health Data.Gov: http://www.healthdata.gov/content/about	Users can search by data category and format (.csv, .xls, zip, PDF, rdf, JSON, html, txt and API)
• Centers for Disease Control and Prevention: http://www.cdc.gov/nchs/data_access/ftp_data.htm	Includes public use files (PUFs) from surveys from multiple government agencies
• Expert Health Data Programming: http://www.ehdp.com/vitalnet/datasets.htm	Host links to about 45 large data sets
• Health Services Research Information Central:https://www.nlm.nih.gov/hsrinfo/datasites.html#488International	Has extensive health datasets, statistics, international data and data tools
• Vanderbilt Biostatistics Datasets: http://biostat.mc.vanderbilt.edu/wiki/Main/DataSets	Multiple health related data sets are available to download as Excel, ASCII, R and S-Plus files. Also includes links to international data sets.
• MIMIC III Critical Care database: https://mimic.physionet.org/	Site is a repository of more than 40,000 de-identified critical care patient-level data
• CMS Data Navigator: https://dnav.cms.gov/Default.aspx	Expedites the search for Medicare and Medicaid data
Free Online Data Science Resources	**Details**
• School of Data: http://schoolofdata.org/courses/	Online course covers data fundamentals, data cleaning, exploring data, extracting data and mapping data
• Class Central: https://www.class-central.com/subject/data-science	Offers multiple free data science and big data-related courses
• Data Science Academy: http://datascienceacademy.com/free-data-science-courses/	Aggregates courses from multiple universities
• Udacity: https://www.udacity.com/courses/data-science	Includes data science courses at beginner through advanced levels
• Oregon Health and Science University (OHSU) Dr. Bill Hersch www.informaticsprofessor.blogspot.com	Free Healthcare Data Analytics Course (9/7/2016) OHSU Big Data to Knowledge (BD2K) Open Educational Resources (OERs) Project (10/23/2016)
• IBM Big Data University: https://bigdatauniversity.com	Offers multiple free courses related to data science and big data analytics for beginners and intermediate level learners
• KDNuggets: www.kdnuggets.com	Check the main "Courses" tab

Free Online Statistics Resources	Details
• Online textbook: http://www-bcf.usc.edu/~gareth/ISL/	An Introduction to Statistical Learning with Applications in R
• Online textbook: http://www-stat.stanford.edu/~tibs/ElemStatLearn/	The Elements of Statistical Learning: Data Mining, Inference, and Prediction, 2nd edition.
• Stat Trek: http://stattrek.com/tutorials/free-online-courses.aspx	Online tutorials guide students through the introductory steps of statistics. There are also brief quizzes and calculators to add interest
• Biostatistics textbook: http://bolt.mph.ufl.edu	Biostatistics. Open Learning Textbook. University of Florida.
• OnlineStatBook: http://onlinestatbook.com/2/index.html and iTunes.apple.com	Excellent Introductory free online reference from David Lane at Rice University. There is also a free e-book for Mac or iOS devices "Introduction to Statistics: An Interactive e-Book.
• OpenIntro: www.OpenIntro.org	Three free PDF download books. One book is associated with about 100 free datasets
• Statistics How To: www.statisticshowto.com	There is an online textbook as well as a companion e-book for sale. The web site includes calculators and stats tables.
• Kaggle: www.kaggle.com	There are forums for those just getting started in data science, as well as information about public data sets. Kaggle also provides job forums for those interested in careers in the data science field. In addition, Kaggle hosts data science competitions for both health and non-health care data.
• StatPages: http://statpages.info/	A mega-site for essentially any free online statistical calculator imaginable.
Spreadsheet Tutorials	**Details**
• University of California at Berkeley: http://multimedia.journalism.berkeley.edu/tutorials/spreadsheets/	Provides the basics on spreadsheets. Based on Google spreadsheets
• Google spreadsheets: https://sites.google.com/a/g.risd.org/training/RISD-Video-Tutorials/google-spreadsheet-tutorials	Good introductory video tutorials on Google spreadsheets
Programming Language Tutorials	**Details**
• Tutorials Point: https://www.tutorialspoint.com/tutorialslibrary.htm	Web site offers tutorials in multiple areas of data science to include the R , Python and SQL Tutorial
• R Tutorial by Kelly Black: http://www.cyclismo.org/tutorial/R/	Interactive introduction to R from the University of Georgia Department of Mathematics
• The Python Tutorial: https://docs.python.org/2/tutorial/	Introductory level instruction on Python
• SQLZoo: http://sqlzoo.net/wiki/SQL_Tutorial	Provides SQL tutorials, references, sample databases
Web Data Extraction Tools	**Details**
• Import .io: https://www.import.io/	Commercial site (free trial available) to scrape data from the web into .csv files
• Google Chrome extension scraper: https://chrome.google.com/webstore/detail/scraper/mbigbapnjcgaffohmbkdlecaccepngid?hl=en	Free Google extension that will convert web based data into a format compatible with Google spreadsheets

Geo-Coding Tools	Details
• Geonames: http://www.geonames.org/	Free international geographical database with over 10 million geographical names, maps, etc.
• QGIS (desktop GIS): http://qgis.org/en/site/about/index.html	Open source desktop GIS tool
Data Science Blogs	**Details**
• KDNuggets: http://www.kdnuggets.com/	Lists 90 blogs that cover most aspects of data science
• Informatics Professor: www.informaticsprofessor.blogspot.com	Dr. Bill Hersh's blog on Biomedical and Health Informatics includes topics related to data science and data analytics and free courses offered by OHSU
Data Science Journals	**Details**
• Data Mining and Knowledge Discovery: http://www.springer.com/computer/database+management+%26+information+retrieval/journal/10618	Six issues published by Springer each year. Available as open access and non-open access
• Data Science Journal: http://datascience.codata.org/	Peer-reviewed open-access journal
• Journal of Data Science: http://www.jds-online.com/	Publishes international research articles on data science. Online access is free

References

1. Data Science Association. http://www.datascienceassn.org/about-data-science Accessed September 16, 2016

2. Data mining. Wikipedia. www.wikipedia.org Accessed January 16, 2016

3. Tukey JW. The future of data analysis. Annals of Math Stats. 1962;33(1):1-67

4. Cleveland WS. Data science: an action plan for expanding the technical areas of the field of statistics. Int Stat Rev 2001. 69(1):21-26

5. Gunelius S. The Data Explosion in 2014 Minute by Minute-Infographic. July 12, 2014. http://aci.info/2014/07/12/the-data-explosion-minute-by-minute-infographic Accessed February 23, 2014

6. Cukier KN, Mayer-Schoenberger V. The Rise of Big Data. Foreign Affairs. May/June 2013. www.foreignaffairs.com Accessed February 15, 2016

7. Press G. A very short history of data science. May 28, 2013. www.Forbes.com Accessed April 15, 2016

8. Google Trends https://www.google.com/trends/ Accessed October 15, 2016

9. Donoho D. 50 years of data science. Presentation at the Tukey Centennial Workshop September 15, 2015. http://courses.csail.mit.edu/18.337/2015/docs/50YearsDataScience.pdf Accessed October 25, 2015

10. Baumer B. A Data Science Course for Undergraduates: Thinking with Data. American Statistician. 2015;69(4):334-342

11. Hersh W. 60 years of Informatics: in the context of data science. Informatics Professor Blog. February 1, 2016. http://informaticsprofessor.blogspot.com Accessed March 1, 2016

12. Stanton J. Chapter 1 About Data. Introduction to Data Science. 2012https://ischool.syr.edu/media/documents/2012/3/DataScienceBook1_1.pdf Accessed January 3, 2016

13. Geraci R. Data analytics: go big or go home. 2015. Business Week 4436, S1-S6

14. White S. Chapter 1 Introduction to Data Analysis in A Practical Approach to Analyzing Healthcare Data. Third edition. AHIMA Press. 2016. Chicago, IL

15. Nonparametric Statistical Methods—3rd Edition. M. Hollander. 2014. John Wiley & Sons. Hoboken, NJ

16. Wikipedia. Normal Distribution. https://en.wikipedia.org/wiki/Normal_distribution Accessed February 25, 2016

17. Standard deviation calculation. Easy calculation.com https://www.easycalculation.com/statistics/standard-deviation.php Accessed February 26, 2016

18. Biostatistics. University of Florida. http://bolt.mph.ufl.edu Accessed March 17, 2016

19. Wilson WF, Meigs JB, Sullivan L, et al. Prediction of incident diabetes mellitus in middle-aged adults. Arch Intern Med 2007;167:1068-1074

20. Hospital Compare. www.medicare.gov/hospitalcompare Accessed February 29, 2016

21. What is a confidence interval? Stat Trek. http://stattrek.com/estimation/confidence-interval.aspx Accessed February 29, 2016

22. Confidence interval for mean calculator. Easy calculation.com https://www.easycalculation.com/statistics/confidence-limits-mean.php Accessed February 29, 2016

23. Kallnowski P. Understanding confidence intervals (CIs) and effect size estimation. Observer. 2010;23(4). www.psychologicalscience.org Accessed February 29, 2016

24. Becker L. Effect size (ES)-effect size calculators. www.uccs.edu/becker/effect-size.html Accessed February 29, 2016

25. Chavalarias D, Wallach JD, Ting Li AH et al. Evolution of reporting p values in the biomedical literature, 1990-2015. JAMA 2016;315(11):1141-1148.

26. Database Primer. www.databaseprimer.com Accessed February 22, 2016

27. Sheta OE, Eldeen AN. The technology of using a data warehouse to support decision-making in health care. Int J Data Man Sys 2013;5(3):75-86

28. Data Mining: Concepts and Techniques. Han J. Elsevier 2000

29. Learn REST: A RESTful Tutorial. http://www.restapitutorial.com Accessed February 25, 2016

30. Huckman R, Uppaluru M. The Untapped Potential of Health Care APIs. Harvard Bus Rev Dec 23, 2015 https://hbr.org/2015/12/the-untapped-potential-of-health-care-apis Accessed February 26, 2016

31. The Argonaut Project. http://hl7.org/fhir/2015Jan/argonauts.html Accessed February 25, 2016

32. Data.Gov API Catalog. http://catalog.data.gov/dataset?q=-aapi+api+OR++res_format%3Aapi#topic=developers_navigation Accessed February 25, 2016

33. Kandel S, Heer J, Plaisant C et al. Research directions in data wrangling: visualizations and transformations for usable and credible data. http://vis.stanford.edu/files/2011-DataWrangling-IVJ.pdf Accessed January 30, 2016

34. Doing Data Science, straight talk from the frontline. O'Neil C, Schutt R. O'Reilly Publisher. 2014

35. Sauro J. 7 Ways to Handle Missing Data. June 2, 2015. http://www.measuringu.com/blog/handle-missing-data.php Accessed January 6, 2016

36. Analytics in a Big Data World: The Essential Guide to Data Science and Its Application. Baesens B. Wiley 2014

37. KD Nuggets. www.kdnuggets.com Accessed January 30, 2016

38. W3 Schools. SQL. http://www.w3schools.com/sql Accessed January 30, 2016

39. Chauhan R. Clustering Techniques: A Comprehensive Study of Various Clustering Techniques. Int J Adv Res Comput Sci 2014;5(5): 97-101

40. WEKA. http://www.cs.waikato.ac.nz/ml/weka/downloading.html Accessed February 25, 2016

41. OnlineStatBook. http://onlinestatbook.com/2/index.html Accessed March 1, 2016

42. Jason Brownlee. Master Machine Algorithms. Discover How They Work and Implement Them from Scratch. 2016. http://machinelearningmastery.com/ Accessed June 4, 2016

43. Yoo I, Alafaireet P, Marinov M et al. Data Mining in Healthcare and Biomedicine: A Survey of the Literature. J Med Syst 2012;36:2431-2448

44. Orange. http://orange.biolab.si Accessed January 3, 2016

45. Peng, R and Matsui E. The Art of Data Science. LeanPress. 2016 https://leanpress.com/ Accessed June 2, 2016

46. Brownlee, J. Machine Learning with WEKA. E-book. 2016. http://machinelearningmastery.com/

47. Vanderbilt Department of Biostatistics. http://biostat.mc.vanderbilt.edu/wiki/Main/DataSets Accessed March 20, 2016

48. IBM Watson Analytics. http://www.ibm.com/analytics/watson-analytics/ Accessed January 4, 2016

49. Nadkarni PM, Ohno-Machado L, Chapman WW. Natural language processing: an introduction. *Journal of the American Medical Informatics Association : JAMIA.* 2011;18(5):544-551. doi:10.1136/amiajnl-2011-000464

50. Monegain B. Natural Language Processing in High Demand. August 14, 2015. Healthcare IT News. http://www.healthcareitnews.com/news/natural-language-processing-demand Accessed August 30, 2016

51. Townsend, H. Natural Language Processing and Clinical Outcomes: The

Promise and Progress of NLP for Improved Care. http://bok.ahima.org/doc?oid=106198 Accessed August 30, 2016

52. Harris B. 5 benefits of natural language understanding for healthcare. Healthcare IT News. October 9, 2012 http://www.healthcareitnews.com/news/5-benefits-natural-language-understanding-healthcare Accessed August 30, 2016

53. Ford E, Carroll JA, Smith HE et al. Extracting information from the text of electronic medical records to improve case detection: a systematic review. JAMIA 2016;23:1007-1015

54. Pennic J. Healthcare Natural Language Processing Market to Reach $2.67B by 2020. HIT Consultant. August 13, 2015. http://hitconsultant.net/2015/08/13/healthcare-natural-language-processing-market-reach-2-67b/ Accessed August 31, 2016

55. Feinerer I. Introduction to the tm Package Text Mining in R. July 3, 2015. https://cran.r-project.org/web/packages/tm/vignettes/tm.pdf Accessed August 31, 2016

56. Ingersoll G. 5 open source tools for taming text. Opensource.com July 8, 2015. https://opensource.com/business/15/7/five-open-source-nlp-tools Accessed September 1, 2016

57. OHNLP Consortium. http://www.ohnlp.org/index.php/Main_Page Accessed September 1, 2016

58. E. R. Tufte. *The Visual Display of Quantitative Information*, 2nd ed., Cheshire, CT, USA: Graphics Press, 2001

59. Evans RS, Benuzillo J, Home BD et al. Automated identification and predictive tools to help identify high-risk heart failure patients: pilot evaluation. JAMIA 2016;23:872-878

60. Microsoft Excel 2016. https://products.office.com/en-us/excel Accessed March 7, 2016

61. Calculating and Reporting Healthcare Statistics. Third Edition. 2010. Horton LA. AHIMA Press.

62. Piktochart. www.piktochart.com Accessed March 7, 2016

63. Marr B. Why only one of the five Vs of big data really matters. March 19, 2015 http://www.ibmbigdatahub.com/blog/why-only-one-5-vs-big-data-really-matters Accessed April 3, 2016

64. What is MapReduce? IBM. https://www-01.ibm.com/software/data/infosphere/hadoop/mapreduce/Accessed June 10, 2016

65. Apache Hive. https://hive.apache.org/Accessed June 10, 2016

66. Apache Mahout. http://mahout.apache.org/Accessed June 15, 2016

67. Raghupathi W, Raguhpathi V. Big data analytics in healthcare: promise and potential. Health Info Sci 2014;2(3). www.hissjournal.com/content/2/1/3 Accessed February 24, 2016

68. NoSQL. http://nosql-database.org/Accessed June 12, 2016

69. Apache Cassandra. http://cassandra.apache.org/Accessed June 10, 2016

70. MongoDB. http://www.mongodb.com/Accessed June 11, 2016

71. Pritchett D. "Base: An Acid Alternative", ACM Queue, vol. 6, no. 3, July 28, 2008

72. Neff G. Why Big Data Won't Cure Us. Big Data. 2013;1(3):117-123 For further reading we refer readers to a 2013 book Frontiers in Massive Data http://www.nap.edu/download.php?record_id=18374# Accessed June 12, 2016

73. Microsoft Excel Analytics ToolPak. https://support.office.com/en-us/article/Use-the-Analysis-ToolPak-to-perform-complex-data-analysis-f77cbd44-fdce-4c4e-872b-898f4c90c007 Accessed March 7, 2016

74. SQL Server Analysis Services. https://msdn.microsoft.com/en-

us/library/hh231701.aspx Accessed March 7, 2016

75. Statistical Package for the Social Sciences (SPSS). http://www-01.ibm.com/software/analytics/spss/ Accessed March 7, 2016

76. IBM Watson Analytics Academic Program. https://www.ibm.com/web/portal/analytics/analyticszone/wanew Accessed January 4, 2016

77. County Health Ranking http://www.countyhealthrankings.org/ Accessed February 20 2016

78. KNIME. https://www.knime.org Accessed February 26, 2016

79. RapidMiner. https://rapidminer.com Accessed February 26, 2016

80. Tableau Public https://public.tableau.com/s/ Accessed January 10, 2016

81. Tableau Whitepaper. Visual Analysis Best Practices http://www.tableau.com/sites/default/files/media/whitepaper_visual-analysis-guidebook_0.pdf Accessed January 10, 2016

82. Qlik Sense Desktop. http://www.qlik.com/products/qlik-sense/desktop Accessed March 8, 2016

83. DataScience Community. http://datascience.community/colleges Accessed March 1, 2016

84. Certified analytics professional. https://www.certifiedanalytics.org Accessed March 2, 2016

85. AHIMA. Certification Health Data Analyst. http://www.ahima.org/certification/chda Accessed March 2, 2016

86. Coursera. www.coursera.org Accessed March 3, 2016

87. Madhani T. University announces $100 million for data science initiative. September 8, 2015 The Michigan Daily. https://www.michigandaily.com/section/news/university-announces-100-million-data-science-initiative Accessed March 1, 2016

88. Indeed. www.indeed.com Accessed March 1, 2016

89. O'Reilly Media Data Science Salary Survey (2015). https://www.oreilly.com/ideas/2015-data-science-salary-survey Accessed February 25, 2016

90. Harpham B. Career Boost: Break into data science. February 25, 2016. www.infoworld.com Accessed February 25, 2016

91. Hersh W. Gimme Some Analytics (We already have it). 2013. Informatics Professor. http://informaticsprofessor.blogspot.com Accessed February 20, 2016

92. McKinsey Global Institute. May 2011. Big data: the next frontier for innovation. http://www.mckinsey.com/business-functions/business-technology/our-insights/big-data-the-next-frontier-for-innovation Accessed February 28, 2016

93. Breiman, L. Statistical Modeling: The Two Cultures. Stat Science. 2001;16(3):199-231

94. Krzysztof JC, Moore GW. Uniqueness of medical data mining. Art Int Med 2002;26:1-24

95. Google Cloud Platform. https://cloud.google.com Accessed June 7, 2016

96. eDX Courses. Microsoft Data Science Professional Project. https://www.edx.org/course/data-science-professional-project-microsoft-dat102x Accessed July 14, 2016

97. National Research Council. Towards Precision Medicine: Building a Network for Biomedical Research and a new Taxonomy of Disease. National Academies Press. 2011 www.nap.edu Accessed October 1, 2016

98. Connecting Health and Care for the Nation. A Shared Nationwide Interoperability Roadmap. October 2015. www.healthit.gov Accessed October 1, 2016

99. Moore's Law. http://www.mooreslaw.org/Accessed June 12, 2016

Chapter 24

Clinical Decision Support

ROBERT E. HOYT

HAROLD P. LEHMANN

Learning Objectives

After reading this chapter the reader should be able to:

- Define electronic clinical decision support (CDS)

- Enumerate the goals and potential benefits of CDS

- Discuss the government and private organizations supporting CDS

- Discuss CDS taxonomy, functionality and interoperability

- List the challenges associated with CDS

- Enumerate CDS implementation steps and lessons learned

"Clinical decision support systems link health observations with health knowledge to influence health choices by clinicians for improved health care."

Robert Hayward, Centre for Health Evidence 2004

Introduction

Definition

The above definition by Dr. Hayward is widely quoted and general in nature. A more recent definition by the Office of the National Coordinator for HIT states "Clinical decision support (CDS) provides clinicians, staff, patients or other individuals with knowledge and person-specific information, intelligently filtered or presented at appropriate times, to enhance health and health care."[1] To be complete, one could argue that any resource that aides the clinician, other healthcare team members or patients with medical decision making should be considered CDS. For the purposes of this chapter we will be only discussing electronic CDS, most commonly that is part of electronic health records (EHRs) and computerized physician order entry (CPOE), discussed in chapter 4 on EHRs. We should emphasize that CDS tools potentially assist more than just physicians; nurses, pharmacists, radiologists, patients, etc. have the same need for knowledge management. Also, we define clinical decision support systems (CDSS) as the information technology systems that support electronic CDS.

Early in its evolution CDS was discussed in terms of alerts and reminders to clinicians, but as pointed out in chapter 4 on EHRs, it should be broadened to include many other tools available to clinicians and others at the point of care. This would include diagnostic help, cost data, up-to-date information about emerging clinical problems such as Zika virus, data from disease registries and so forth. The vision is for all CDS data to be electronic, structured and computable, which is frequently not the case.

In spite of extensive online medical resources and robust search engines available to all members of the healthcare team, questions concerning correct diagnoses and optimal treatments still arise frequently.[2] For that reason, many experts have strongly promoted CDS as a driving force towards improved patient care and safety. Therefore, CDS and quality are closely related.

The Five Rights of CDS

CDS subject matter experts have stressed that in order for CDS to be effective it must include five rights:

1. The right information (what): should be based on the highest level of evidence possible and adequately referenced. There should be good internal and external validity; that is, the recommendations are based on high quality studies and they can be applied to a similar population of patients. These issues are discussed in the chapter 14 on evidence based medicine and clinical practice guidelines.

2. To the right person (who): that is the person who is making the clinical decision, be that the physician, the patient or some other member of the healthcare team.

3. In the right format (how): should the information appear as part of an alert, reminder, infobutton or order set? That depends on the issue and clinical setting.

4. Through the right channel (where): should the information be available as an EHR alert, a text message, email alert, etc.?

5. At the right time (when) in workflow: new information, particularly in the format of an alert should appear early in the order entry process so clinicians are aware of an issue before they complete, e.g. an electronic prescription. This adds to the usability of the technology.[3]

Historical Perspective

Interest in clinical decision support is not new. In 1959 Ledley and Lusted published an article on medical reasoning in which they predicted computers would be used for medical decision making.[4] Early experience with CDSSs began in the 1970s with several well-known initiatives. These employed a variety of systems and knowledge bases that will be discussed in more detail in a later section. With the exception of the HELP system, these programs were standalone programs, not integrated with any other systems, such as the EHR and most are no longer available.

De Dombal's system for acute abdominal pain: was developed at Leeds University in the UK in the 1970s. It was based on Bayesian probability to assist in the differential diagnosis of abdominal pain.[5]

Internist-1: was a diagnostic CDS program developed at the University of Pittsburg in the 1970s that relied on production rules (IF-THEN statements). It used patient observations to generate possible diagnoses. In spite of being cumbersome, the knowledge base was used in the subsequent system known as QMR.[6]

Mycin: was a rule-based system developed by Dr. Edward (Ted) Shortliffe and others at Stanford University in the 1970s to diagnose and treat infections.[7]

Later, in the 1980s several initiatives achieved some commercial success.

DxPlain: was developed in 1984 by Massachusetts General Hospital as both a medical reference system and diagnostic CDSS. Based on clinical findings (signs, symptoms, laboratory data) the program generated a ranked list of diagnoses related to the clinical manifestations. DxPlain justifies the diagnoses, suggests further steps and describes atypical manifestations. As a reference, it describes over 2400 diseases, the signs and symptoms, causes and prognosis, as well as references. Institutions can lease the program annually.[8]

QMR: Quick Medical Reference was a diagnostic CDDS consisting of an extensive knowledge base of diagnoses, symptoms and lab findings. It evolved from Internist-1 in the 1980s and discontinued about 2001.[9]

HELP: Health Evaluation Through Logical Processing (HELP) was developed in the 1980s by the University of Utah as a hospital information system and later as a diagnostic CDSS. It provides alerts and reminders, data interpretation, diagnostic help, management suggestions and clinical practice guidelines. They developed the Antibiotic Assistant that was later commercialized into TheraDoc.™ It is still in use today in Intermountain Healthcare hospitals as part of their EHR system. It has been modernized and is now known as HELP2. They have developed thousands of medical logic modules (MLMs) used for CDS and discussed later in the chapter.[10]

Iliad: is a diagnostic CDSS and reference system for professionals that was developed by the University of Utah in the 1980s and discontinued about 2000. The last version (CD-ROM) covered about 930 diseases and 1500 syndromes with ICD-9 codes for each diagnosis.[11-12]

An extensive list of early initiatives is archived on the web site OpenClinical.[13] The following are a sample of more recent CDS tools:

Isabel: created in 2002, is a differential diagnosis web-based tool available for worldwide use. Signs and symptoms are inputted as free text or imported from the EHR. and a diagnostic checklist is generated as a standalone tool or integrated with the EHR (SNOMED-CT coded). The inference engine relies on natural language processing with a database of about 100,000 documents and 40 proprietary algorithms. It is offered as a paid subscription and a mobile app is available.[14]

SimulConsult: is a diagnostic program based on Bayesian networks with its strength in pediatrics, genetics, neurology and rheumatology. For the knowledge base it includes about 5,500 diagnoses, and 3000+ genetic variants. A search is conducted using the patient's age and gender and then symptoms, signs, lab tests, images, etc. are added to narrow the differential diagnosis. The program also includes recommended additional testing and articles to read. It is particularly useful for a child with genetic variants and unusual physical findings where a differential diagnosis is important.[15]

SnapDx: a free mobile app that is a diagnostic CDS for clinicians. It is based on positive and negative likelihood ratios (LRs) derived from the medical literature. As of mid-2016 the program covered about 50 common medical conditions (Apple iOS only).[16] (See figure 24.1)

Figure 24.1 SnapDx on smartphone

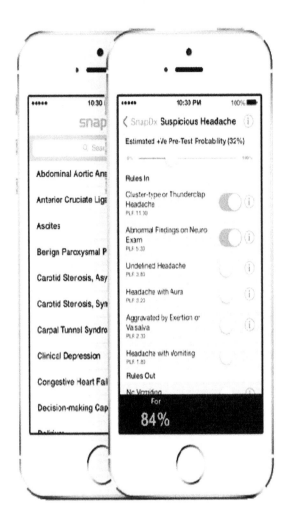

CDS Benefits and Goals

Electronic health records alone are electronic filing cabinets of healthcare data. Without additional tools they could also be referred to as electronic data management systems (EDMSs). It is when they include helpful information and guidance that they have the potential to positively impact patient care.

Some of the potential benefits and goals of CDS based on several expert sources are listed in Table 24.1. [3, 17-18]

Table 24.1 CDS Benefits and Goals

Benefits and Goals	Details
Improvement in patient safety	• Medication alerts • Improved ordering
Improvement in patient care	• Improved patient outcomes • Better chronic disease management • Alerts for critical lab values, drug interactions and allergies • Improved quality adjusted life years (QALY) • Improved diagnostic accuracy
Reduction in healthcare costs	• Fewer duplicate lab tests and images • Fewer unnecessary tests ordered • Avoidance of Medicare penalties for readmission for certain conditions • Fewer medical errors • Increased use of generic drugs • Reduced malpractice • Better utilization of blood products
Dissemination of expert knowledge	• Sharing of best evidence • Education of all staff, students and patients
Management of complex clinical issues	• Use of clinical practice guidelines, smart forms and order sets • Interdisciplinary sharing of information • Case management
Monitoring clinical details	• Reminders for preventive services • Tracking of diseases and referrals
Improvement of population health	• Identification of high-cost/needs patients • Mass customized messaging
Management of administrative complexity	• Supports coding, authorization, referrals and care management
Support clinical research	• Ability to identify prospective research subjects

Organizational Proponents of Clinical Decision Support

Multiple governmental and non-governmental organizations have been proponents of CDS. The following are known leaders in the field.

Institute of Medicine (IOM)

The IOM has been a strong proponent of the use of information technologies to improve access to clinical information and support clinical decision making. In Crossing the Quality Chasm, they stated "*Development of decision support tools to assist clinicians and patients in applying t he evidence, evidence - based care processes, supported by automated clinical information and decision support systems, offer the greatest promise of achieving the best outcomes from care for chronic conditions.*" In many of their subsequent publications they also mentioned CDS as an evolutionary tool.[19]

American Medical Informatics Association (AMIA)

In 2006 A Roadmap for National Action on CDS was published under the auspices of AMIA. The Roadmap outlined three pillars of CDS:

1. *"Best Knowledge Available When Needed: the best available clinical knowledge is well organized, accessible to all, and written, stored and transmitted in a format that makes it easy to build and deploy CDS interventions that deliver the knowledge into the decision making process*

2. *High Adoption and Effective Use: CDS tools are widely implemented, extensively used, and produce significant clinical value while making financial and operational sense to their end-users and purchasers*

3. *Continuous Improvement of Knowledge and CDS Methods: both CDS interventions and clinical knowledge undergo continuous improvement based on feedback, experience, and data that are easy to aggregate, assess, and apply. "*

AMIA has also established a CDS Working Group to carry out the vision of the Roadmap and to offer a forum for the advancement of CDS.[20-21]

Office of the National Coordinator (ONC)

ONC funded research by the RAND Corporation and Partners Health Care/ Harvard Medical School in a project known as "Advancing Clinical Decision Support". Four tasks have been outlined:

- *"Task 1: Distill best practices for CDS design and CDS implementation, preparing resources on best practices for broad dissemination through a variety of online channels.*

- *Task 2: Distill best practices and standards for sharing CDS knowledge and produce an open online platform for sharing CDS knowledge artifacts (such as alerts, order sets, etc.) among EHR vendors and/or provider organizations.*

- *Task 3: Develop a "clinically important" drug-drug interaction (DDI) list, as well as a legal brief about the liability implications of using the clinically important DDI list.*

- *Task 4: Develop a process that engages specialty bodies in weighing performance gaps vs. CDS opportunities to select targets for meaningful use of CDS by specialists"*

ONC also has a special section on CDS sharing which is important if CDS is expected to become widespread and vetted by multiple organizations. Sharing is intended to be part of Task 4 above and should include these activities:

- *"Identified key requirements and features for a Knowledge Sharing Service (KSS)*

- *Proposed important attributes of a governance model and editorial policy*

- *Proposed an architecture based on leveraging existing standards where available*

- *Proposed standard XML schemas for commonly deployed intervention types, which could be imported into a vendor system*

- *Identified gaps in standards that present barriers to progress and make recommendations to appropriate standards development organizations*

- *Deployed a first-generation KSS*

- *Populated the KSS with illustrative XML CDS interventions targeting Meaningful Use and implemented custom style sheets so the XML files could be viewed in human-readable form"*

ONC established the Health eDecisions (HeD) project which ran from 2012-2014 to develop and harmonize CDS standards. They developed an XML schema for representing knowledge and a Virtual Medical Record (vMR) for representing patient data. In 2014 ONC launched the Clinical Quality Framework Initiative to further harmonize those standards that would integrate guidelines for electronic clinical quality measures (eCQMs) and CDS. ONC has also been instrumental in the development of the Quality Data Model that includes clinical concepts necessary for eCQMs, as part of the meaningful use program.[22-23]

Agency for Healthcare Research and Quality (AHRQ)

AHRQ funded several CDS initiatives that began in 2008:

- The Clinical Decision Support Consortium (CDSC) is an initiative to develop and test web-based delivery of CDS, primarily for EHRs. The goal of the CDS Consortium is *"to assess, define, demonstrate, and evaluate best practices for knowledge management and clinical decision support in healthcare information technology at scale – across multiple ambulatory care settings and EHR technology platforms."* [24]

Access to newly developed CDSs is through the public CDSC Knowledge Management Portal that supports search and retrieval of CDS interventions. CDS can be unstructured or highly structured with encoded logic. Brigham and Women's Hospital was the major partner with participating organizations located throughout the US. As of early 2015 the portal contained over 100 CDS tools consisting of: alerts, reminders, templates, order sets, reference information, Infobuttons, and value sets. The tools can be viewed in XML format or in human-readable style sheets. In mid-2014 the Consortium was placed on hiatus, awaiting additional funding.

- Guidelines into Decision Support (GLIDES). Yale School of Medicine was charged with translating CPGs into structured data for the outpatient treatment of common diseases. (Example: childhood obesity and asthma). Guidelines would be disseminated at multiple sites using different EHRs. This generated the Guideline Elements Model (GEM), a knowledge model for guidelines that incorporates a set of more than 100 tags to categorize guideline content.

- CDS eRecommendations project used 45 recommendations from the US Preventive Services Task Force and 12 recommendations relevant to stage 1 meaningful use and created a framework for use with EHRs. Their final report was published in 2011.

- CDS Key Resources: Two CDS white papers and other monographs are included on the AHRQ web site and are available in the resource section of this chapter.

- US Health Information Knowledgebase (USHIK) is an AHRQ initiative to support knowledge "artifacts" using HL7 Health eDecision (HeD) schema. The XML versions can be downloaded and shared by healthcare organizations or EHR vendors seeking to meet meaningful use CDS requirements. [24-26]

Health Level 7 (HL7)

This international standards development organization (SDO) has a working group dedicated to advancing electronic CDS with the following goals:

- *"Work on CDS standards for knowledge representation, such as, Infobuttons and order sets*
- *Work on patient-centered monitoring such as alerts and reminders*
- *Work on population-centric monitoring and management, such as disease surveillance*
- *Work on representation of CPGs*
- *Develop a data model for clinical decision support*
- *Identify existing HL7 messages and triggers for CDS"*

HL7 has also developed the Fast Healthcare Interoperability Resources (FHIR) standard that can be used for CDS and will be discussed in another section. [27]

National Quality Forum (NQF)

The NQF CDS Expert Panel met in 2010 and developed a CDS taxonomy with four components: triggers, input data, intervention and action steps. Their goal was to create the taxonomy for future quality performance measures and to map to the quality data set (QDS) model, necessary for electronic clinical quality measures (eCQMs). [28]

Leapfrog

This patient safety organization has long promoted computerized physician order entry (CPOE). As part of their approach they developed a CPOE Evaluation Tool that tests a hospital's EHR with multiple mock scenarios, to include 12 CDS categories, such as therapeutic duplications. [29-30]

Healthcare Information and Management Systems Society(HIMSS)

HIMSS published one of the best known textbooks on CDS implementation. [2] Moreover, they created the Electronic Medical Record Adoption Model (EMRAM) that rates the different levels of US EHR adoption from 1 to 7. Full use of CDS would qualify as a stage 6 level.

By early 2016, of the participating healthcare organizations 29.1% of hospitals had achieved level 6, but only 14.3% by ambulatory facilities.[31]

Centers for Medicare and Medicaid Services (CMS) Meaningful Use Program

CMS is responsible for reimbursing eligible physicians and hospitals for meaningful use of certified EHRs. CMS views clinical decision support to be integral to quality measures and the improvement of patient care. In stage 1 the following objectives were related to CDS:

- Capturing clinical data in a standard, coded manner.

- Utilizing computerized provider order entry.

- Implementing drug-drug, drug-allergy, and drug-formulary checks.

- Setting patient reminders per patient preference.

- Performing medication, problem, and medication allergy reconciliation at transitions of care.

Stage 2 meaningful use required greater use of CDS tools. In 2014, eligible professionals had to report on 9 out of 64 total CQMs, while eligible hospitals and critical access hospitals had to report on 16 out of 29 total CQMs. Specifically, to comply with MU core measure 6 they were required to do the following:

- *"Implement five CDS interventions related to four or more CQMs, if applicable, at a relevant point in patient care for the entire EHR reporting period"*

- *"Enable the functionality for drug-drug and drug-allergy interaction checks for the entire EHR reporting period"*[32]

In spite of support from multiple agencies and organizations, widespread implementation of CDSSs has been slow. This will be discussed in more detail in the section on CDS Challenges.

CDS Methodology

CDS entails two phases: Knowledge Use and Knowledge Management.

Knowledge Use involves several typical steps demonstrated in figure 24.2 which is the taxonomy created by the National Quality Forum. Triggers are an event such as an order for a medication by a user. Input data refers to information within, for example the EHR, that might include patient allergies. Interventions are the CDS actions such as displayed alerts. The action step might be overriding the alert or canceling an order for a drug to which the patient is allergic.[33] Managing this knowledge-based process is called *knowledge engineering*. Because most decision support is performed through tools and processes set by vendors, most knowledge engineers do not call themselves by this explicit name, but perform the tasks implicitly.

Figure 24.2 CDS Use Phases

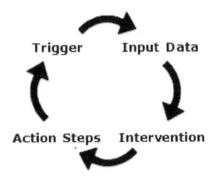

Most of these decision support systems are table driven or depend on software coding that embeds clinical, regulatory, or administrative knowledge into the programming driving the system. Focusing on the technology underlying CDSSs, Berner articulates 2 high-level classes where the knowledge is separable from the clinical-system programming: knowledge based (KB) and non-knowledge based (NKB).[34] While systems based on these architectures (algorithms) are uncommon in clinical operations, we expect them to become more prevalent as the clinical-data environment stabilizes and developers figure out how best to use these more advanced methods in the operational environment.

Knowledge Management, in turn, involves knowledge acquisition, knowledge representation, and knowledge maintenance.

Knowledge Acquisition includes expert-based knowledge or data-based knowledge. The former may come from clinical practice guidelines external to the organization and from clinical expertise from within the organization. Data-based knowledge may also come from models built on data from outside the institution (e.g., APACHE scoring [35]) or from data mining from within the institution. Specific methods are described shortly.

Knowledge Representation has many forms. The choice of representation should depend on the problem at hand, the level of expertise of the knowledge engineer, the resources available, and the commitment of the institution to knowledge maintenance. The classical knowledge-based CDS consists of a knowledge base (evidence based information), an inference engine (software to integrate the knowledge with patient-specific data) and a means to communicate the information to the end-user, such as a pop-up alert in the EHR. There are several different types of knowledge representations in these KB CDSSs:

- **Configuration**: Here, knowledge is represented by the choices specified by the institution, which may be context sensitive.

  Thus, a formulary may "know" the insurance status of a patient. Knowledge acquisition involves having the committee in charge of the clinical system (e.g., pharmacy) involved, and knowledge maintenance, reviewing those choices on some frequency or as problems arise.

- **Table based**: Many systems represent their rules for care in tables. For instance, drug–drug interactions have one basic rule: If the patient is on one drug in a row in the table, and an order is placed for a second drug in the same row, and the level of alerting is above the threshold specified by the institution, then the ordering clinician will be alerted. The rows of drug pairs, along with their level of danger, represent drug-drug–interaction knowledge. Knowledge may be acquired from a vendor (in the case of drug–drug interactions) or from the CDS Committee. Knowledge maintenance involves vendor updates and periodic reviews, in light of patient-quality indicators or patient-safety events.

- **Rules based**: also known as production rules and "expert systems". The knowledge base consists of IF-THEN statements, such as, if the patient is allergic to sulfa derivatives and a sulfa drug is prescribed, then an alert will be triggered.[7] In contrast to tables, the

rules may be of arbitrary complexity. The attraction of rules is that they should be *modular,* in that any single rule can be debugged in isolation of others. The classic example of rules-based CDS is MYCIN, which, while demonstrating many features necessary for CDSs, was not adopted into practice, primarily because the system could not have its needed in-use data fed to it by the host EHR.[36] The rule-based architecture is preferable to code-driven software, because to modify the latter means to modify software that directly runs the clinical system. Knowledge acquisition here is generally internal; efforts to share rules have often floundered on the problem that MYCIN faced: how to get data of specific EHRs of different environments into the same rules.[37] See **CDS Standards**, below.

- **Bayesian networks**: These structures use forms of Bayes' Theorem (conditional probabilities) to calculate the (posterior) probabilities of diseases (or other state of concern), based on the pretest probability, prevalence of each disease, P(Disease), conditioned on patient-specific data (such as symptoms). For example, the probability of a disease given a positive test using Bayes' Formula would be:

$$P(Disease|Test+) = \frac{P(Test+|Disease) \times P(Disease)}{P(Test+)}$$

P(Disease|Test+) (also called positive predictive value) and P(Test+|Disease) (also called sensitivity) are conditional probabilities. To deal with multiple findings (tests, signs, symptoms), two assumptions are often made; each finding is independent of other findings and the findings can be grouped into categories, such as present or absent. The results can be both surprising and enlightening.[38] (See example posted by Cornell University in the infobox). Bayesian systems in use involve networks of diseases and findings; knowledge acquisition means to specify the many conditional probabilities in the network. The odds form of Bayes theorem is Posterior Odds (Disease|Test Positive) = Prior Odds (Disease) x Likelihood Ratio Positive; the log form of this has the log prior odds added to the log likelihood ratio (LLR); the LLR then has the semantics of "weight of

Bayesian Probability Example:

The prevalence of breast cancer is about 1% in women (ages 40-50) and a woman with breast cancer has a 90% chance of a positive test (mammogram). In addition, there is a 10% chance of a false positive test (mammogram is positive but woman doesn't have cancer).

Based on this information, what is the probability that a lady age 45 has breast cancer based on a positive mammogram?

Answer: 9 in 108 or about 8% [38]

evidence." A variety of free Bayesian calculators are available on the Internet. Examples of Bayesian CDS are: Iliad, Gideon, SimulConsult and De Dombal's system.[5,11,15,39] Systems that are pseudo-Bayesian (that is, use weighing of evidence in non-Bayesian ways) include Dxplain.[8]

Knowledge maintenance means there is a need to keep knowledge up to date, from the level of the program through the committees in charge and to track changes and reasons. This maintenance has proven challenging to most organizations. [40]

Non-Knowledge based CDS

When the knowledge representation is a model derived from data mining, the system is called a non-knowledge based CDS. Data mining methods may involve artificial intelligence (AI), (e.g. neural networks, machine learning) or more traditional statistical methods, like linear or logistic regression. These are *data-based* systems, that require the models be developed and validated prior to being used in clinical operation. An open-source program for all these methods is the WEKA environment. [40] The power of these approaches is that the methods first can analyze large data sets, looking for new trends or patterns at the population level. Then, the resulting model can derive recommendations specific to the patient at hand, and lie at the heart of Predictive Analytics. [41] AI machine learning has also been used for "pattern recognition" which has become relatively routine in medical diagnostic devices (interpreting images, electrocardiograms, etc.). This approach does not require patient symptoms or physiological findings for the interpretation. For AI models that do require such data, there is no agreement in the field about the amount of validation that is needed to incorporate a model

developed elsewhere into a local system. Data mining and predictive modeling can be categorized as supervised or unsupervised machine learning.

Supervised machine learning assumes that the user knows ahead of time what classes or categories exist. A training sample is used that has target (dependent) variables and input (independent) variables. With multiple training sessions the goal is to narrow the gap between observed and expected observations. The predictive model can be further classified as a classification model if the target involves discrete (categorical or nominal) data (can only be certain values, such as blood type) and a regression model if the target involves continuous data (data that can be anything numerical in a range, such as a patient's weight).

An important aspect of supervised learning is classification or the mapping of data into predefined classes. Techniques to accomplish this are:

- The computational model known as **neural networks** was first reported in 1943 by McCulloch and Pitts. [40] Neural networks are a popular approach and are capable of both supervised and unsupervised machine learning. The networks are arranged in layers and each layer is an array of processing elements or neurons. The input layer receives multiple inputs and in the hidden layer signals are processed and an output is generated to the output layer. Using the supervised model, the outputs are compared to the target output and training with input-output pairs is repeated until the trained output and desired target output are similar. (See figure 24.3). The results are dependent on a good training set, otherwise the results may not reflect a larger population.[41]

Figure 24.3 Neural networks

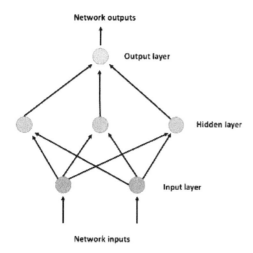

- **Logistic regression** is used to analyze data where the output is binary, (for example, the patient has or doesn't have hypertension). The independent variables or input is computed against the output/outcome and odds ratios are produced. The end result is a prediction model and this approach has been used for many years by healthcare researchers.[42]

- **Decision trees**. There are two types of "decision trees" used in knowledge engineering. One is a *classification tree*,

where the attributes of an individual case (patient, population, setting) are input, and the algorithm classifies to one class or the other. Typical algorithms here are C4.5 and recursive tree partitioning, both available in many statistical and machine learning packages. [41,46-47]

The "tree" here has to do with use of the attributes, generally in a binary way and recursively (i.e., the same operation is done at each step). The other type of decision tree is used in *decision analysis* to derive an optimal action, or event, a flowchart of action.[48] The "tree" here lays out the space of possible outcomes, which have more to do with unfolding over time than with attributes. They generally consist of decision nodes (squares), chance nodes (circles) and terminal nodes or outcomes (triangles). Probabilities are assigned to the path branches, while costs or other measures of value or preference are assigned to the outcomes. Figure 24.4 displays a decision tree for deciding if patients need a soft, hard or no contact lens. In this example, the decision tree calculates that the first logical branch is tear production. These decision trees are a subset of the wider set of *decision models*. While mostly used to inform policy, a few decision models have been implemented to provide patients real-time decision support as part of *shared decision making*.[49]

Figure 24.4 Decision Tree for Contact Lens Recommendations

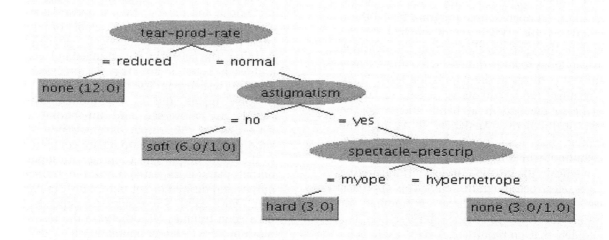

Unsupervised machine learning analyzes data without any classes so the learning system develops new classes or patterns (clustering). This is a valuable approach if the user desires an exploratory data analyses, not based on prior classifications. There are several different types of unsupervised learning.

- **Cluster analysis** is a technique that organizes a large data set into distinct groups. Most commonly they are grouped as a hierarchy. Cluster analysis has been used to group gene sequences, for example identifying clusters of genes associated with breast cancer. Although this technique is fast, it is challenged by clusters that may be difficult to interpret. [50] An example of how this approach is used can be found in an article by Newcomer et al. that evaluated a cohort of Kaiser-Permanente patients that were in the top 20% for cost of care over two years. This approach identified discrete groups of chronically ill patients that would benefit from care management. [51] This is one of the many approaches healthcare organizations, particularly accountable care organizations, might adopt to lower costs.[52]

Other machine learning approaches include:

- Support Vector Machines
- Learned Bayesian Networks
- Association Rules

CDS Standards

Much medical knowledge is independent of the specific institution where it is used: calcium and bicarbonate should *never* be in the same IV line, because limestone will precipitate in the line and in the vein; patients surviving myocardial infarction should be treated with beta-blockers, because the best evidence shows a decrease in subsequent mortality.[53] Patients with type 1 diabetes do better if they have close control; however, implementing this directive depends on local resources. Thus, for several decades, developers of CDS have struggled with how to share the medical knowledge that rises above local concerns, while enabling local committees to modify the rules to be in accord with local realities. The following are some of the significant standards developed and certified to address these concerns.

Arden Syntax: this standard was developed in 1999 and is now an ANSI/HL7 standard (v2.7-2008). The core representations are Medical Logic Modules (MLMs), which encode information for a single medical decision (text file), so it can be shared. There are a standard set of categories (*maintenance, library, knowledge*) and slots (e.g., within *knowledge, data* and *logic*). It is an open standard that can be used by individuals, healthcare organizations and vendors for development of clinical rules. The problem is that MLMs can't be shared and this is known as the "curly braces problem" because the assignment of local data to the rules variables, in the *data* slot, depends on institution-specific programming code placed between curly braces ({}) in that slot. Note in figure 24.5 hematocrit is contained within a pair of curly braces. In spite of the fact that several EHR vendors use this standard, overall usage has been low. [54-55]

Figure 24.5 Data slot of an MLM [56]

```
data:
        blood count storage := event
                ('complete blood count');
        hematocrit := read last (('hematocrit'):
        previous_hct := read last (('hematocrit'))
                where it occurred before the
                time of hematocrit);;
```
A patient's hematocrit may be stored as "hematocrit" in the EHR, or as "Hct" or as a database SELECT instruction. The specifics comprise the content of the curly braces.

GELLO: is a class-based object oriented language that can create queries to extract and manage data from EHRs, in order to create decision criteria. GELLO is an attempt to address the curly braces problem by inputting the patient data required from a "virtual EMR," which a vendor would then be responsible for linking to. It was also developed by HL7 as a standard query and expression language. It is part of HL7 version 3 and provides the framework for manipulation of clinical data for CDS.[57]

GEM: is an international ASTM standard developed so clinical practice guidelines, the source of much of the knowledge, could be stored and shared in a XML format. GEM addresses the lack of uniformity on the source of knowledge, in particular, that clinical guidelines are written as text, which requires much interpretation to turn into machine-directing rules. Yale University developed this standard and the GLIDES initiative discussed in an earlier section. Another feature is the "GEM Cutter", an XML editor that facilitates guideline markup.[58]

Guideline Interchange Format (GLIF): was developed by several biomedical informatics programs to enable sharable and computer-interpretable guidelines at the knowledge level.[59]

Clinical Quality Language (CQL): The quality-improvement community discovered that population-level imperatives look very similar to clinically-directed rules. CQL is a draft HL7 standard being evaluated as a new language to represent eCQMs and CDS. It facilitates human and machine readable representations in XML.[60]

Infobuttons: Infobuttons refers to the function of linking patient data to general information; they are often implemented as Web-linked icons that permit downloading of guidelines, articles, monograph entries, or other canonical information. Infobuttons are associated with an HL7 Infobutton Management Standard that standardizes context-sensitive links embedded within EHR systems. The OpenInfobutton project is an open-source initiative by the Veterans Health Administration (VHA) and the University of Utah. As an example, the content provider UpToDate connects ICD-9 codes, lab results and medication information of individual patients via EHR infobuttons to their online knowledge base.[61-62]

Fast Healthcare Interoperability Resources (FHIR): is a draft HL7 standard that holds great promise for healthcare sharing. The JASON Task Force that was part of the HIT Standards and Policy Committee recommended that FHIR be used and public application programming interfaces (APIs) be adopted by EHR vendors to promote interoperability. FHIR is a RESTful API that is a http based standard that uses either XML or JSON for data representation, OAuth for authorization and ATOM for queries. This same web services approach is used by Facebook and Google. FHIR is data and not document centric so EHR (A) can place a http request for data from EHR (B); a clinician could request just the results for one lab test, instead of a consolidated CDA document. This approach would also facilitate interactions between an EHR and a CDS database/server. In addition, this strategy permits apps to be created that will interface with EHRs. Figure 24.6 demonstrates an actual CDS app for monitoring bilirubin levels in infants used by Intermountain Healthcare.

Figure 24.6 Smart app on FHIR

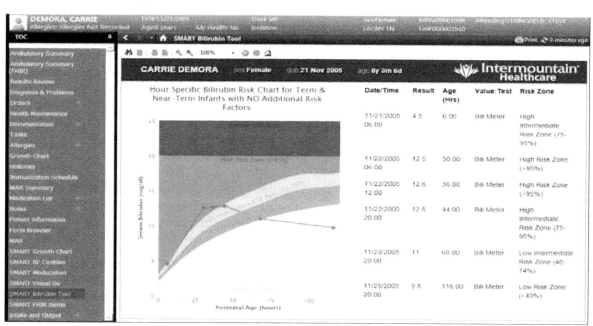

The Argonaut Project was created to promote and test this strategy with leaders in the healthcare and EHR vendor industries. HL7 has created a variety of clinical and administrative "resources" (structured data) that can be requested. Examples of resources might be a medication lists, allergies, and problem lists and these can be combined in a message. Each resource would use available standards such as LOINC, RxNorm, etc. This approach is faster, simpler and more flexible than other attempts at interoperability.

CDS is one of the FHIR resources, such that a query can be initiated for a decision based on parameters. For example, FHIR could be used to identify a drug-drug interaction.[27, 63-64] As of mid-2016 twenty-three SMART apps are listed in the app gallery.[64]

CDS Functionality

There are several other ways to classify CDSSs, in addition to knowledge based and non-knowledge based systems. For example, Osheroff et al. developed a taxonomy of functional intervention types that was related to the type and timing of tasks.[3]

Importantly, CDS can be located within the EHR or link to a remote CDS program. CDS can activate before, during or after a patient encounter. Activation can be automatic or "on demand." Alerts can be non-interruptive or interruptive (e.g. the process stops until the physician responds to the alert).

Table 24.2 lists a taxonomy of CDS that is based on CDS goals, stated earlier. There is overlap among the categories, as sections such as patient safety and patient care are interrelated.[3, 17-18]

Wright and colleagues and the Provider Order Entry Team (POET) at Oregon Health & Science University elicited a taxonomy from 11 CDS experts, and validated the taxonomy with 9 vendors and 4 CDSS developers.[65] Based on function, they arrived at 3 high-level categories: Order facilitators, point-of-care alerts/reminders, and relevant information displays.

Ordering Facilitators. While prescribing support is the main focus of order entry CDS, it is important to realize that clinicians (and patients) need guidance regarding appropriate lab and imaging orders, as well as cost data. With healthcare reimbursement heading in the direction of value based purchasing, there needs to be greater emphasis on ordering the drugs,

labs and images of choice that are evidence based (more comments on this in chapter 4 on EHRs).

Order sets: are special orders in EHRs that are customized to provide guidance for common problems encountered with hospital admissions. As an example, for community acquired pneumonia (CAP) an order set might recommend a certain antibiotic, when the follow-up chest X-ray will be ordered, oxygen use, etc. This is an effort to standardize care based on evidence based medicine. Order sets are researched, adopted and modified by local clinicians.

Order sets should make the process of writing admission orders faster with choices provided with drop down lists. The order set standard is an HL7 standard that aims to standardize order set libraries.[66] There has not been a great deal of research in this area. An earlier report by the Veterans Administration (VA) and a report in 2010 summarized their experience with order sets in seven disparate healthcare systems. Both studies confirmed that creating and maintaining order sets is labor intensive, but they save time and improve care. They also point out that many of the order sets were underutilized so ongoing scrutiny is needed.[67-68]

There are commercial providers of order sets, such as Zynx that offers more than 800 order sets capable of being integrated with many EHR vendors.[69]

Therapeutic support: programs such as Theradoc® provide a clinical surveillance program that includes an *Infection Control Assistant* (monitors hospital acquired infections) and *Pharmacy Assistant* that provides drug surveillance of antibiotics and anticoagulants (blood thinners). This platform integrates with clinical EHR data.[70] It is not known how many commercial EHR vendors offer clinical calculators for common medical issues and risk prediction. Allscripts EHR has fully integrated a battery of calculators (68) in their workflow.[71] Calculators can be purchased or home-grown. For example, the Cleveland Clinic wanted to improve prescribing the blood thinner (heparin) so they created an online calculator that was eventually integrated into their EHR.[72] The new SMART platform with FHIR will make it possible to develop apps that integrate with commercial EHRs with open APIs in the near future. A variety of CDS therapeutic tools should become available.[64]

Table 24.2 CDS Taxonomy

Function	Examples
Patient Safety	• Medication alerts • Critical lab alerts • Ventilator support alerts • Improved drug ordering for warfarin and glucose • Infusion pump alerts • Risk calculation • Improved legibility • Diagnostic aids
Cost	• Reminders to use generic drugs or formulary recommendations • Fewer duplications • Reminders about costs of drugs, lab tests and imaging studies • Reduce Medicare penalties for readmissions • Reduced medication errors • Reduced malpractice claims • Better utilization of blood products
Patient Care	• Embedded clinical practice guidelines, order sets, and clinical pathways • Better chronic disease management • Identify gaps in recommended care • Immunization aids • Diagnostic aids • Sepsis alerts (**see Case Study infobox**) • Antibiotic duration alerts • Prognostic aids • Patient reminders • Pattern recognition for images, pulmonary function tests and EKGs, blood gases, Pap smear interpretation
Disseminating Expert Knowledge	• Use of infobuttons for clinician and patient education • Provide evidence based medicine with embedded clinical practice guidelines and order sets
Managing Complex Clinical Issues	• Reminders for preventive care for chronic diseases • Care management • Predictive modeling based on demographics, cost, and clinical parameters
Managing Complex Administrative Issues	• Decision modeling • Research recruitment

---

Case Study: Prediction of ICU sepsis using minimal EHR data

Researchers used machine learning software (InSight) to predict sepsis based on a validated ICU data set (MIMIC-III) of 40,000+ patients. The predictors were: vital signs, oxygen saturation, Glasgow Coma Score and age; all easily obtainable from most EHRs.

Using the test dataset with a prevalence of sepsis of 11.3% machine learning classification performance was compared with the existing sepsis scoring systems SIRS, SOFA and MEWS. The results using receiver operating characteristic (ROC) curves and area under the curve (AUC) measures, InSight outperformed the other risk scoring systems. Additionally, they randomly deleted 60% of the data and still outperformed the other scoring systems.

It should be noted that this was a retrospective and not prospective study, so further research will be necessary to see if eventual patient morbidity and mortality is reduced by utilizing this machine learning model in the ICU setting

Desautels T, Calvert J, Hoffmann J et al. Prediction of Sepsis in the Intensive Care Unit With Minimal Electronic Health Record Data: A Machine Learning Approach. JMIR 2016;4(3):e28

Smart forms or smart templates: are electronic encounter notes that include structured questions pertinent to selected diseases, augmented with CDS. For example, a smart form for diabetes would include questions that center on care of the diabetic patient with the usual co-morbidities. The advantage of this approach is that it is relatively fast and the output is primarily structured data. Also, the smart form can have recommended diagnostic tests and treatments embedded in the form. Partners HealthCare has been a pioneer in smart forms and has designed them for acute respiratory tract infections, diabetes and coronary artery disease. One of the downsides of this approach is that it generates a "robotic" note that does not provide a historical narrative. It is unknown how many templates combine a structured approach with free text (hybrid approach). Templates are generally used for very mild common illnesses such as sore throat, disability determinations, etc. For a discussion of the issues related to structured notes versus free text narratives, see the article by Rosenbloom et al.[73]

Alerts and Reminders: alerts serve to warn the clinician about potential problems, during or after the patient visit. Common alerts appear as pop-up boxes and are triggered by electronic prescribing, ordering of labs or images and return of results. The alerts can be interruptive, forcing the clinician to provide a response or non-interruptive, providing information or advice during a patient encounter. Reminders, as the name implies, remind a clinician or patient about preventive medicine measures

such as "mammogram is due". These are often triggered before the patient is seen and may or may not appear as pop-up boxes.

Relevant Information Displays:

- **Infobuttons, hyperlinks, mouse-overs:** infobuttons were covered in the section on CDS standards. These tools can be used by any member of the healthcare team or patients. Depending on the EHR vendor, users can connect to a variety of embedded internal and external educational resources. For example, the medical resource UpToDate offers an embedded link for nine major commercial EHRs.[74]

- **Diagnostic support:** non-integrated differential diagnosis support is available through Simulconsult and DxPlain.[8,15] Isabel, discussed earlier, is capable of automatically pulling coded symptom and sign data from the EHR.[14] Ultimately, natural-language-processing programs will be more routine, able to pull in all types of data and fully integrate with the EHR, so that support tools are auto-populated with demographics and clinical data, thus saving the clinician time.

- **Dashboards:** are an infrequently discussed type of CDS, because they are aimed at *population-based* decision making, rather than individual clinical decision making. Defined broadly, dashboards can include any type of patient summary, flow chart of lab results or vital signs or a disease registry. Excellent data representation helps the clinician track and trend common medical problems.

CDS Sharing

Currently, there is no single way that CDS knowledge is represented and shared. Importantly, there doesn't seem to be a universal incentive for disparate healthcare organizations to share CDS tools and other HIT. Various CDS standards exist as discussed in a prior section. Ideally, standards would allow for sharing among healthcare organizations and EHR vendors, but in 2016, we do not have CDS interoperability. Thus far, two approaches have been taken: structure the knowledge base for sharing, using a standard vocabulary or access a central repository of CDS using service oriented architecture (SOA).[3] Both ONC and AHRQ have sponsored CDS sharing initiatives in the past, as outlined in the prior chapter section *Clinical Decision Support by Organizations*.

Several other CDS initiatives should be mentioned. The Socratic Grid is an open-source SOA platform that supports CDS of clinical data and utilizes more than just production rules and offers a variety of prediction models.[75] OpenCDS is another open-source open-standards SOA platform that provides CDS tools and resources. This public-private collaboration uses open patient data as input and conclusions/recommendations are delivered as output.[76]

ONC convened an expert panel on CDS sharing in 2013 and they listed the key principles for sharing: "(1) prioritize and support the creation and maintenance of a national CDS knowledge sharing framework; (2) facilitate the development of high-value content and tooling, preferably in an open-source manner; (3) accelerate the development or licensing of required, pragmatic standards; (4) acknowledge and address medicolegal liability concerns; and (5) establish a self-sustaining business model".[77]

It seems likely FHIR will become a new approved standard for healthcare data sharing, including CDS, in the near future.

CDS Implementation

Multiple excellent resources are available for CDS implementation. Table 24.3 will outline some of the logical steps used for the management of most CDS implementation projects.

CDS Challenges

General

Robert Greenes aptly stated "the difficulty in deploying and disseminating CDS is in large part due to the lack of recognition of how hard the job is and lack of availability of tools and resources to make this job easier." Furthermore, not all CDS is the same; diagnostic CDS is more complicated than simple allergy alerts.[81]

CDS is clearly challenged by the knowledge explosion in medicine, big data from digitized medicine and the addition of genomics into clinical practice. Currently, incentives to use CDS is aligned with meaningful use payments but this program does not apply to all physicians and what happens when the program goes away?

Organizational support: Clearly, for CDSSs to be successful there has to be support from the administrative, IT and clinical leadership. It should be part of the mission and vision of the healthcare organization. This implies that CDS is a priority, there are incentives and it is achievable. Only large healthcare organizations with a track record of innovation and advanced IT support can create and maintain a sophisticated CDDS, as we currently know them. It is still not clear that CDS has enough value or a strong business case as it currently stands.

CDS Reports and Reviews: Much has been written about the impact of CDS on medical and administrative processes, as well as patient outcomes. This section will summarize several of the major reviews on this topic. It should be noted that there have been many reviews of the impact of health information technology (HIT) on medical and administrative processes, as well as reviews on computerized physician order entry (CPOE) that includes CDS. Moreover, many reviews were written before the HITECH Act and the EHR Incentive program, such that studies performed before the widespread adoption of EHRs may have different results from the post-adoption period.[82] The following are reviews that specifically evaluate CDSSs:

- Two systematic reviews by Garg et al. and Kawamoto et al. in 2005 discussed early CDS impact. According to Garg et al. 64% of the

Table 24.3 CDS Implementation Steps

Logical Steps	Details
Project initiation	• Ensure clinical and non-clinical leadership are onboard and have a shared vision • Ensure CDS is synched with organizational goals, patient safety/quality measures and meaningful use objectives • Determine the business case/value of CDS for the organization • Determine feasibility from a manpower and financial standpoint and acceptance by clinicians • Ensure objectives are clear and attainable • Identify key stakeholders and assess buy-in • Understand that the CDS needs of specialists are different from primary care • Assess readiness, EHR capability and IT support • Assess the clinical information systems (CISs) involved • Assess knowledge management capabilities • Assemble the CDS team: clinical leaders, CMIO, administrative and nursing leaders, managers, EHR vendor and IT experts • Identify clinical champions • Develop CDS charter
Project planning	• Consider a SWOT analysis (strengths, weaknesses, opportunities and threats) • Utilize standard planning tools such as Gantt charts and swim lanes • Develop timeline • Decide whether to build or buy CDS content • CDS committee should select CDS interventions that fit their vision • Be sure to follow the 5 Rights of CDS • Map the different processes involved with CDS and be sure they integrate with the clinician's workflow • Determine whether you will measure structure, processes and/or outcomes • Plan the intervention: triggers, knowledge base, inference engine and communication means • Educate staff and gain their input • Design the CDS program for improvement over baseline performance in an important area for the organization. In other words, be sure you can measure outcomes and compare with baseline data • Investigate the needed CDS standards required • Follow the mandates of change management, e.g. John Kotter's Eight Step Model • Communicate goals of CDS project to all affected
Project execution	• Provide adequate training and make CDS training part of EHR training • Develop use cases • Test and re-test the technology: unit, integration and user acceptance testing • Decide on incremental roll-out or "big bang" • Provide a mechanism for feedback in the CDS process, as well as formal support
Project monitoring and control	• Use data from feedback, override logs, etc. to modify the system as needed • Compare the alert and override rates with national statistics • Measure percent of alerts that accomplished desired goals • Communicate the benefits and challenges to the end-users as they arise • Use tools such as the AHRQ Health IT Evaluation ToolKit • Knowledge management maintenance; are guidelines unambiguous and up to date? Who will maintain the content?

The implementation specifics provided came from the following references.[4,23,78-80] studies showed improvement in clinician performance and 13% showed improvement in actual patient outcomes.[83] Kawamoto et al. similarly found that 68% of the studies demonstrated improvement in clinical practice. They identified the following success factors: CDS was integrated into clinical workflow; CDS was electronic, not paper based; CDS provided support at the right time and location and CDS provided actionable recommendations.[84]

- Berner wrote a CDS State of the Art paper for AHRQ in 2009. She made the point that many studies focused on inpatients and many were done at academic medical centers with home grown EHRs and strong IT support.

  This may yield different results from studies based on commercial EHRs in community settings. She also noted that many of the studies were not randomized controlled trials and most focused on processes and not patient outcomes. Additionally, there were methodological limitations noted in most studies. The point was made that qualitative studies are also needed to better understand why some CDSS implementations work and others do not.[34]

- Black et al reported in 2011 on the impact of health information technology (eHealth) on healthcare quality and safety. Like others, this effort focused on systematic reviews from 1997 to 2010. Overall, they felt the studies were of poor quality. CDS was a category of their review but they concluded that there was a large gap between the potential and proven benefits of all HIT, to include CDSSs. Evaluations tended to focus on benefits with little attention to unintended adverse consequences and cost. The authors included a table of studies demonstrating CDS benefit and a table for evidence of risks.[85]

- Lobach et al. published an evidence report on CDS and knowledge management for AHRQ in 2012. Their review covered the period 1976 through 2010, to include 148 randomized controlled trials. They concluded that commercially and locally developed CDSSs (inpatient and outpatient) improved preventive services, ordering clinical studies and prescribing therapies. They identified six new success factors: integration with CPOE, promotion of action, lack of need of additional data entry by clinicians, evidence

based CDS, local involvement and CDS for patients and clinicians. Only 20% of trials reported on actual patient outcomes and only 15% reported on cost.[86]

- Jasper et al. reported a systematic review on prior CDS systematic reviews (SRs) in 2011. They used the AMSTAR rating system for systematic reviews and only examined SRs with a score of 9 or above. In spite of using this scoring system, only 17 out of 35 SRs were of high quality. They concluded that 57% of SRs showed practitioner improvement and 30% showed a positive impact on patient outcomes. They made the point that simple CDS, such as drug alerts and preventive care reminders are likely successful because they require minimal patient data.[87]

- Bright et al., in another systematic review of the literature from 1976 to 2011, reported their results in 2012. The authors were the same as in the review by Lobach. Like other reviews they noted the majority reported on the effect of CDS on processes and not outcomes or cost. They commented that few mentioned unintended adverse consequences, workflow or efficiency issues and studies were heterogeneous, making comparisons difficult.[88]

- Jones et al. reported a systematic review of HIT, with a focus on meaningful use covering the time period from 2010 to 2013. Fifty-seven percent of the reports evaluated CDS, with 65 percent being positive, 17 mixed, 11 neutral and 7 negative. The authors concluded that most reports were positive in regards to the effects on quality, safety and efficiency, but there was a paucity of information about contextual and implementation factors that would have been helpful to determine why some implementations were not successful.[89]

To summarize, there is ample evidence to show CDS improves a variety of processes but there is much less evidence to show a positive impact on clinical outcomes, such as mortality, length of stay, health-related quality of life and adverse events. The same holds for measuring economic outcomes.[4]

Unintended adverse consequences (UACs): As with all HIT implementations, unanticipated and undesirable side effects are discovered over time.   Ash et al. convened an expert panel to analyze UACs related to CDSSs. They concluded that there were two major UAC patterns noted: problems associated with CDS content and the presentation of information on the computer screen. Content issues were related to shifting of human roles, outdated CDS and misleading CDS. Presentation issues were related to rigidity, alert fatigue and a variety of potential errors such as incorrect auto-completes and timing issues.[90]

Alert fatigue is perhaps the most publicized UAC related to CDSSs. van Der Sijs et al. performed a review of this issue in 2006 with several important observations. Physicians override alerts between 49-96% of the time but most are for good reasons. Only 2-3% of the time does overriding result in a true adverse drug event (ADE). Reviewers tended to agree with the overrides 95.6% of the time. These data suggest that too many alerts are unnecessary and poorly written. [91] High override rates (81-87%) were also reported out of the VA system, however a study in Boston showed an acceptance rate of 67% for interruptive alerts.[92-93] Clearly, how well the alerts are written, how they appear in workflow and how well the innovation is accepted plays a role.

Medico-legal: because CDSSs make recommendations that may alter patient outcomes, they have legal ramifications. CDS needs to be based on the best recognized evidence and must be updated regularly. Ignoring EHR generated alerts or treatment recommendations also has legal implications. This raises the question, do healthcare organizations need to archive alerts and responses, in case there are negative outcomes? [94] For the above reasons the FDA has suggested that they regulate HIT that has patient safety implications. As of early 2016, the FDA has not provided guidance regarding CDS regulations and this may have a dampening effect on CDS innovation and development. [95]

In addition, product liability for CDS malfunctioning may discourage innovation by EHR vendors.

Clinical: CDS must fit into the workflow of all members of the healthcare team. CDS needs to be integrated within the EHR and adhere to the five rights of CDS, covered earlier in the chapter. Currently, most CDS is part of meaningful use objectives, but to be truly successful, other decisions must be supported, shared, updated and measured. For example, medication CDS

must eventually include adjustments for the patient's weight, gender, renal and liver function, mental status, age, etc.

Technical: as discussed in other sections, uniform standards and an acceptable interoperability platform are necessary so that CDS can be standardized, updated, shared and stored. Also, in order for CDS to be well received and promulgated it must be well designed and match usability criteria. Furthermore, the creation of local CDS (not vendor generated) is complicated and labor intensive, requiring a multi-disciplinary team, coupled with maintenance and evaluation. CDS is therefore an example of a "socio-eco-technical" implementation.

CDS will likely be more complicated in the future as we move beyond meaningful use objectives and consider incorporating genetic data into decision support. Highly complex CDS will magnify alert fatigue and the need for near constant updating of rules and algorithms.

Lack of interoperability: the lack of interoperability has been one of the most challenging obstacles to EHRs and HIT in general. Ideally, CDS should be shared so that healthcare organizations do not have to reproduce work that has already been done. As discussed in prior sections, multiple other organizations are working on a solution.

Long term CDS benefits: Durability remains a question mark. Multiple studies have shown that as long as there is scrutiny of a systematic change, like CDS, there seems to be improvement; but when the oversight diminishes, so do the results.[96]

Lessons Learned

There is a wealth of literature on avoiding the pitfalls of CDS design, implementation and maintenance. We will summarize some of the best known lessons learned and recommendations from a variety of resources into Table 24.4. [4, 34, 70-71, 97-98]

Many of the studies were conducted pre-HITECH Act so better data may be available as CDS interventions are generated and reported as part of meaningful use. For example, the study by Roshanov et al. suggested that standalone CDS programs were linked to a higher success rate than those what were part of CPOE.[99] Will this hold up with newer data? We also don't

Table 24.4 Lessons Learned

Lesson Learned	Comments
Project initiation • Healthcare organizations have competing priorities • CDS cannot come from external mandate	Ensure the organization can support a new CDS initiative. Even if CDS is intended to match meaningful use, it must be embraced by all and match organizational goals
Project planning • Customization of content and workflow is important • One size CDS does not fit all • CDS must match the 5 Rights of CDS • Make CDS as non-intrusive and non-interruptive as possible • Ideally, there should be recommendations for clinicians and patients • Interventions should include a reason for overrides • Intervention should make recommendation and not just assessment • "Do CDS with users, not to them" • EHR data must be up to date for triggers to work correctly	Customization is desirable but labor intensive and not available at smaller organizations. Specialists and primary care clinicians have different needs. Clinicians do not want to stop and speed is important. Table 24.1 CDS Taxonomy
Project execution • Feedback buttons in CDS work well • Include CDS training into EHR training • CDS must be tested for UACs and patient safety	User feedback is critical
Project monitoring and control • Knowledge management is time consuming • Be sure intervention content is up to date	There may have to be a separate knowledge management committee

know if CDS content that is purchased will be a better or worse choice for organizations.

Executing CDS the way it comes out of the box is not enough. Given that alert fatigue is so prevalent (up to 95% ignoring alerts), implementers have the obligation of looking at the entire decision making/therapeutic process and not just the physician's interaction with the system.

Recommended Resources

Textbooks

• Berner ES. Clinical decision support systems theory and practice. Clinical decision support systems theory and practice. 2007.[100]

- Greenes Robert A. Clinical Decision Support: The Road to Broad Adoption. Second Edition. 2014[81]

- Osheroff Improving Outcomes with Clinical Decision Support. An Implementer's Guide. Second Edition. 2012 HIMSS. Chicago, IL[4]

Web sites

- AHRQ CDS Initiatives [26]

- ONC CDS Resources [23]

White papers

- Lobach et al. Enabling Health Care Decision making Through Clinical Decision Support and Knowledge Management: Evidence Rep. 2012 [86]

- Berner E. Clinical Decision Support Systems: State of the Art (PDF) 2009 [34]

- Karsh BT. Practice Improvement and Redesign: How Change in Workflow Can Be Supported by CDS (PDF). AHRQ 2009.[101]

Future Trends

In the section on CDS sharing it was reported that subject matter experts felt that the US needed a national CDS knowledge sharing framework. Presumably, this would be the responsibility of ONC or AHRQ. To accomplish this, there would likely need to be more data standards created and a business case for widespread CDS adoption. Clearly, Meaningful Use objectives provide a business case for the near future. The proposed Stage 3 Meaningful Use includes the following objectives related to CDS for eligible physicians:

1) implement five CDS interventions related to four quality measures

2) enable drug-drug and drug-allergy interaction alerts for the entire EHR reporting period.[102]

However, it is unclear what the business case will be after that. As the experts also point out, the liability issues will need to be worked out.[94]

A newer direction appears to be using FHIR for interoperability and the creation of CDS applications that integrate with commercial EHRs. Using this approach there would be new incentives for developers to create a variety of decision support tools for clinicians and patients.

Key Points

- With evolving technologies and strategies, the definition of CDS should be broad
- There is widespread support for CDS implementation as part of meaningful use
- In spite of widespread support, CDS faces multiple challenges
- The evidence, thus far, suggests CDS benefits processes more than patient outcomes

Conclusion

Busy clinicians often do not incorporate the best evidence into their medical practices and the hope is that robust CDS, that is part of modern EHRs, will offer them the best evidence within their natural workflow.[101] The Institute of Medicine (IOM) has promoted the idea of the "learning health system" that is part of US healthcare of the future. The third recommendation towards reform promoted

"Decision support tools and knowledge management systems should be routine features of healthcare delivery to ensure that decisions made by clinicians and patients are informed by current best evidence."[103]

The concept of clinical decision support is not in its infancy, but the use of electronic CDS that is fully integrated with EHRs is new. Therefore, we still need better research as to what works and what does not. To date, attempts to deliver the best evidence via CDS has produced mixed results. It is likely we will see further innovations arising that will make CDS more usable and more acceptable by the average busy clinician.

References

1. Clinical Decision Support. Office of the National Coordinator. http://www.healthit.gov/policy-researchers-implementers/clinical-decision-support-cds   Accessed January 15, 2015

2. Ely JW, Osheroff JA, Maviglia SM et al. Patient-Care Questions That Physicians Are Unable to Answer. JAMIA 2007;14(4):407-414

3. Ledley RS, Lusted LB. Reasoning Foundations of Medical Diagnosis. Science. 1959;130(3366):9-21

4. Improving Outcomes with Clinical Decision Support: An Implementer's Guide Jerome A. Osheroff, Jonathan M. Teich, Donald Levick, Luis Saldana, Ferdinand T. Velasco, Dean F. Sittig, Kendall M. Rogers, Robert A. Jenders: Second Edition. 2012 HIMSS. Chicago, IL

5. De Dombal FT, Leaper DJ, Staniland JR, McCann AP, and Horrocks JC. Computer-aided diagnosis of acute abdominal pain. British medical journal. 1972;2(5804):9.HL7 http://www.hl7.org/Special/committees/dss/overview.cfm Accessed February 8, 2015

6. Parker RC, and Miller RA. Creation of realistic appearing simulated patient cases using the INTERNIST-1/QMR knowledge base and interrelationship properties of manifestations. Methods Inf Med. 1989;28(4):346-51. Data Model update Jan 2015 http://www.healthit.gov/sites/default/files/qdm_4_1_2.pdf

7. Shortliffe EH, Davis R, Axline SG, Buchanan BG, Green CC, and Cohen SN. Computer-based consultations in clinical therapeutics: Explanation and rule acquisition capabilities of the MYCIN system. Computers and Biomedical Research. 1975;8(4):303-320. doi:10.1016/0010-4809(75)90009-9.

8. DxPlain http://www.mghlcs.org/projects/dxplain Accessed February 12, 2015

9. Quick Medical Reference (QMR) http://www.openclinical.org/aisp_qmr.html Accessed February 15, 2015

10. Health Evaluation Through Logical Processing (HELP) http://intermountainhealthcare.org/isannualreport/2009/Website/help2.html

11. Iliad 4.5 http://www.ramex.com/title.asp?id=1292 Accessed February 9, 2015

12. Illiad http://www.openclinical.org/aisp_iliad.html Accessed February 9, 2015

13. OpenClinical http://www.openclinical.org/aisinpractice.html Accessed February 9, 2015

14. Isabel http://www.isabelhealthcare.com/home/default Accessed February 9, 2015

15. SimulConsult www.simulconsult.com Accessed February 1, 2015

16. SnapDx http://www.snapdx.co/ Accessed February 9, 2015

17. Perreault LE, Metzler JB. A pragmatic framework for understanding clinical decision support. Journal of Healthcare Information Management. 1999;13(2):5-21.

18. Gilmer TP, O'Connor PJ, Sperl-Hillen JM, Rush WA, Johnson PE, Amundson GH, Asche SE, and Ekstrom HL. Cost-effectiveness of an electronic medical record based clinical decision support system. Health Serv Res. 2012;47(6):2137-58. doi:10.1111/j.1475-6773.2012.01427.x.

19. Corrigan JM. Crossing the quality chasm. Building a Better Delivery System. 2005.

20. Osheroff JA, Teich JM, Middleton B, Steen EB, Wright A, and Detmer DE. A roadmap for national action on clinical decision support. Journal of the American Medical Informatics Association: JAMIA. 2009;14(2):141-5. doi:10.1197/jamia.M2334 Available at: http://www.pubmedcentral.nih.gov/articlerender.fcgi?artid=2213467&tool=pmcentrez&rendertype=abstract

21. American Medical Informatics Association (AMIA) www.amia.org Accessed January 28, 2015

22. Advancing Clinical Decision Support. RAND corporation. http://www.rand.org/health/projects/clinical-decision-support.html Accessed January 29, 2015

23. Office of the National Coordinator. Clinical Decision Support. http://www.healthit.gov/policy-researchers-implementers/clinical-decision-support-cds Accessed January 29, 2015

24. CDS Portal http://cdsportal.partners.org CDSCSearch.aspx Accessed January 29, 2015

25. CDS Consortium http://www.cdsconsortium.org/Accessed February 1, 2015

26. Clinical Decision Support Initiatives http://healthit.ahrq.gov/ahrq-funded-projects/clinical-decision-support-cds-initiative Accessed January 9, 2015

27. HL7 www.hl7.org Accessed January 8, 2015

28. Quality Forum. Driving Quality and Performance Measurement—A Foundation for Clinical Decision Support: A Consensus Report. NQF. Washington DC 2010. http://www.qualityforum.org/Publications/2010/12/Driving_Quality_and_Performance_Measurement_-_A_Foundation_for_Clinical_Decision_Support.aspx Accessed January 8, 2015

29. Kilbridge PM, Welebob EM, and Classen DC. Development of the Leapfrog methodology for evaluating hospital implemented inpatient computerized physician order entry systems. Qual Saf Health Care. 2006;15(2):81-4. doi:10.1136/qshc.2005.014969.

30. Leapfrog Group. www.leapfroggroup.org Accessed January 29, 2015

31. HIMSS Electronic Medical Record Adoption Model http://www.himssanalytics.org/emram/emram.aspx Accessed January 29, 2015

32. Centers for Medicare and Medicaid Services. www.cms.gov Accessed January 29, 2015

33. Driving Quality and Performance Measurement—Foundation for Clinical Decision Support. NQF.

http://www.qualityforum.org/Publications/2010/12/Driving_Quality_and_Performance_Measurement__A_Foundation_for_Clinical_Decision_Support.aspx. Accessed February 28, 2015

34. Berner ES. Clinical decision support systems: state of the art. AHRQ Publication. 2009;(09-0069):4-26.

35. Wong DT, Knaus WA. Predicting outcome in critical care: the current status of the APACHE prognostic scoring system. Can J Anaesth 1991; 38:374-83

36. Buchanan BG, Shortliffe EH. Rule-Based Expert Systems: The MYCIN Experiments of the Stanford Heuristic Programming Project. Reading, MA: Addison-Wesley; 1984

37. Introduction to Knowledge Systems. Mark Stefik. Morgan Kaufmann Publishing. 1995. San Francisco, CA.

38. Baye's Formula. Cornell University. http://www.math.cornell.edu/~mec/2008-2009/TianyiZheng/Bayes.html Accessed February 10, 2015

39. Gideon. http://www.gideononline.com/ Accessed July 25, 2016

40. Geissbuhler A, Miller RA. Distributing knowledge maintenance for clinical decision support systems: the knowledge library model. Proc AMIA Symp 1999:770-774

41. WEKA Machine Learning. http://www.cs.waikato.ac.nz/ml/weka/ Accessed July 25, 2016

42. Predictive Analytics. http://www.predictiveanalyticstoday.com/what-is-predictive-analytics/ Accessed July 26, 2016

43. McCulloch WS, and Pitts W. A logical calculus of the ideas immanent in nervous activity. The bulletin of mathematical biophysics. 1943;5(4):115-133.

44. Sordo M. Introduction to Neural Networks i Healthcare. 2002. Open Clinical http://www.openclinical.org/docs/int/neuralnetworks011.pdf Accessed February 2, 2015

45. What is logistic regression? Statistics Solutions. http://www.statisticssolutions.com/what-

is-logistic-regression/ Accessed July 26, 2016

46. C4.5 Programs for machine learning. J. Ross Quinlan. 1993. Morgan Kaufmann Publishers, San Mateo, CA.

47. Wang F. Adaptive semi-supervised recursive tree partitioning: The ART towards large scale patient indexing in personalized healthcare J. Bio Inform 2015; 55:41-54

48. Hunink MGM. Decision Making in Health and Medicine: Integrating Evidence and Values. Cambridge; Cambridge University Press; 2001

49. Gambhir SS, Shepherd JE, Shah BD et al. Analytical Decision Model for the Cost Effective Management of Solitary Pulmonary Nodules. J Clin Onc 1998;16(6):2113-2125

50. Hardin JM, Chhieng DC. Data mining and clinical decision support systems. Chapter 3 in Berner ES. Clinical decision support systems theory and practice. 2007. Springer. Second edition.

51. Newcomer SR, Steiner JF, and Bayliss EA. Identifying subgroups of complex patients with cluster analysis. The American journal of managed care. 2010;17(8): e324-32.

52. Colak C, Karaman E, Turtay MG et al. Application of knowledge discovery processes on the prediction of stroke. Compute Methods Programs Biomed 2015 119(3):181-185

53. AHRQ Beta-Blockers for Acute Myocardial Infarction. http://archive.ahrq.gov/clinic/commitfact.htm Accessed February 8, 2016

54. Arden Syntax. HL7. http://www.hl7.org/implement/standards/product_brief.cfm?product_id=2 Accessed February 1, 2015

55. Jenders RA. Decision Rules and Expressions. Chapter 15 in Greenes RA. Clinical decision support the road to broad adoption. 2014.

56. Hripcsak G. Writing Arden Syntax Medical Logic Modules. Comput Biol Med. 1994 Sep;24(5):331-63.

57. GELLO. HL7. http://www.hl7.org/implement/standards/product_brief.cfm?product_id=5 Accessed February 2, 2015

58. GEM. ASTM. http://www.astm.org/Standards/E2210.htm Accessed February 3, 2015

59. GLIF. Biomedical Informatics Research and Development Center. University of Rochester. http://www.birdlab.org/research-glif.cfm Accessed February 2, 2015

60. Strasberg H. Clinical Quality Language. www.solutions.wolterskluwer.com January 12, 2015 Accessed March 1, 2015

61. OpenButton Project http://www.openinfobutton.org/ Accessed February 3, 2015

62. UpToDate. Infobutton. http://www.uptodate.com/home/hl7 Accessed February 3, 2015

63. FHIR HL7. http://hl7-fhir.github.io/index.html Accessed February 10, 2016

64. SMART on FHIR http://smarthealthit.org/smart-on-fhir/ Accessed June 2, 2016

65. Wright A, Sittig DF, Ash JS et al. Development and evaluation of a comprehensive CDS taxonomy: comparison of front-end tools in commercial and internally developed electronic health record systems. J Am Med Assoc 2011;18(3):232-242

66. HL7 Order Set Standard. http://www.hl7.org/implement/standards/product_brief.cfm?product_id=287 Accessed March 8, 2015

67. Payne T, Hoey P, Nichol P et al. Preparation and Use of Pre-constructed Orders, Order Sets and Order Menus in a Computerized Provider Order Entry System. Journal of the American Medical Informatics Association. 2003;10(4):322-329

68. Wright A, Sittig EF, Carpenter JD et al. Order Sets in Computerized Physician Order Entry Systems: An Analysis of Seven Sites. AMIA 2010 Symposium Proceedings. :892-896

69. Zynx Health. www.zynxhealth.com Accessed February 7, 2016

70. Theradoc. www.theradoc.com Accessed March 21, 2015

71. eCalcs. http://www.galenhealthcare.com/products-services/products/ecalcs/ Accessed March 6, 2015

72. Butterfield S. Let the computer do the math. ACP Hospitalist August 2014. www.acphospitalist.org Accessed March 7, 2015

73. Rosenbloom ST, Denny JC, Xu H, Lorenzi N, Stead WW, and Johnson KB. Data from clinical notes: a perspective on the tension between structure and flexible documentation. Journal of the American Medical Informatics Association. 2011;18(2):181-186

74. UpToDate www.uptodate.com Accessed March 4, 2015

75. Socratic Grid www.socraticgrid.org Accessed March 24, 2015

76. Open CDS. www.opencds.org Accessed March 24, 2015

77. Kawamoto K, Hongsermeier T, Wright A, Lewis J, Bell DS, and Middleton B. Key principles for a national clinical decision support knowledge sharing framework: synthesis of insights from leading subject matter experts. Journal of the American Medical Informatics Association. 2013;20(1):199-207 http://www.ncbi.nlm.nih.gov/pmc/articles/PMC3555314/

78. Project Management www.project-management.com Accessed March 24, 2015

79. Byrne C, Sherry D, Mercincavage L et al. Advancing Clinical Decision Support. Key Lessons in Clinical Decision Support Implementation. Westat Technical report. 2011. http://www.healthit.gov/sites/default/files/acds-lessons-in-cds-implementation-deliverablev2.pdf Accessed January 20, 2015

80. Kotter J. Leading change. Boston Mass.: Harvard Business School Press; 1996.

81. Greenes RA. Clinical decision support. The road to broad adoption. 2014. Second edition. Elsevier.

82. Romano MJ, Stafford RS. Electronic Health Records and Clinical Decision Support Systems. Impact on Ambulatory Care Quality. Arch Int Med 2011; 10:897-903

83. Garg AX, Adhikari NK, McDonald H, Rosas-Arellano MP, Devereaux PJ, Beyene J, Sam J, and Haynes RB. Effects of computerized clinical decision support systems on practitioner performance and patient outcomes: a systematic review. JAMA. 2005;293(10):1223-38. doi:10.1001/jama.293.10.1223.

84. Kawamoto K, Houlihan CA, Balas EA, and Lobach DF. Improving clinical practice using clinical decision support systems: a systematic review of trials to identify features critical to success. Bmj. 2005 ;330(7494) :765. doi:http://dx.doi.org.

85. Black AD, Car J, Pagliari C, et al. The Impact of eHealth on the Quality and Safety of Health Care: A Systematic Overview. PLoS Medicine. January 2011; 8(1). www.plosmedicine.org

86. Lobach D, Sanders GD, Bright TJ, Wong A, Dhurjati R, Bristow E, Bastian L, Coeytaux R, Samsa G, and Hasselblad V. Enabling health care decision making through clinical decision support and knowledge management. 2012

87. Jaspers MWM, Smeulers M, Vermeulen H, and Peute LW. Effects of clinical decision-support systems on practitioner performance and patient outcomes: a synthesis of high-quality systematic review findings. Journal of the American Medical Informatics Association: JAMIA. 2011;18(3):327-334.

88. Bright TJ, Wong A, Dhurjati R, Bristow E, Bastian L, Coeytaux RR, Samsa G, Hasselblad V, Williams JW, Musty MD, Wing L, Kendrick AS, Sanders GD, and Lobach D. Effect of clinical decision-support systems: a systematic review. Ann Intern Med. 2012;157(1):29-43. doi :10.7326/0003-4819-157-1-201207030-00450

89. Jones SS, Rudin R, Perry T et al. Health Information Technology : An Updated Systematics Review with Focus on Meaningful Use. Ann Intern Med. 2014 ;160(1) :48-54

90. Ash J. S., Sittig, D. F., Campbell, E. M., Guappone, K. P., & Dykstra, R. H. (2007). Some unintended consequences of clinical decision support systems. AMIA … Annual Symposium Proceedings / AMIA Symposium. AMIA Symposium, 26–30.

91. Van der Sijs H, Aarts J, Vulto A, Berg M. Overriding of drug safety alerts in computerized physician order entry. JAMIA 2006; 13:138-147.

92. Lin C-P, Payne TH, Nichol WP, Hoey PJ, Anderson CL, and Gennari JH. Evaluating clinical decision support systems: monitoring CPOE order check override rates in the Department of Veterans Affairs' Computerized Patient Record System. Journal of the American Medical Informatics Association. 2008;15(5):620-626

93. Shah NR, Seger AC, Seger DL, Fiskio JM, Kuperman GJ, Blumenfeld B, Recklet EG, Bates DW, and Gandhi TK. Improving acceptance of computerized prescribing alerts in ambulatory care. Journal of the American Medical Informatics Association. 2006;13(1):5-11.

94. Greenberg M, and Ridgely MS. Clinical decision support and malpractice risk. JAMA: the journal of the American Medical Association. 2011;306(1):90-91

95. Wicklund E. FDA urged to clarify clinical decision support regulations. mHealth Intelligence. February 26, 2016. www.mhealthintelligence.com Accessed September 16, 2016

96. Gerber JS, Prasad PA, Fiks AG et al. Durability of benefits of an outpatient antimicrobial stewardship intervention after discontinuation of audit and feedback. Research Letter. December 17 2014 JAMA;2014312(23):2569-2570

97. Bates DW, Kuperman GJ, Wang S et al. Ten Commandments for Effective Clinical Decision Support: Making the Practice of Evidence-based Medicine a Reality. Journal of the Amer Med Info Assoc 2003;10(6):523-530

98. Eichner J and Das M. Challenges and Barriers to Clinical Decision Support Design and Implementation Experienced in the Agency for Healthcare Research and Quality CDS Demonstrations. AHRQ Pub. No.10-0064-EF March 2010

99. Roshanov PS, Fernandes N, Wilczynski JM et al. Features of effective computerized clinical decision support systems: meta-regression of 162 randomized trials. BMJ 2013;346: f657 doi: 10.1136/bmj. f657

100. Clinical Decision Support Systems. Theory and Practice. Second Edition. Eta S. Berner, Editor. 2007. Springer. New York, NY. http://www.unimasr.net/ums/upload/files/2012/Mar/UniMasr.com_ad22bb3650b5a1fa7e31e56a8e03f3a0.pdf

101. Karsh BT. Practice Improvement and Redesign: How Change in Workflow Can Be Supported by CDS (PDF). AHRQ 2009 http://healthit.ahrq.gov/sites/default/files/docs/page/09-0054-EF-Updated_0.pdf Accessed January 5, 2016

102. Stage 3 Proposed Rule https://s3.amazonaws.com/public-inspection.federalregister.gov/2015-06685.pdf Accessed June 2, 2016

103. Best Care at Lower Price: The Path to Continuously Learning Health Care in America. 2012 http://www.iom.edu/Reports/2012/Best-Care-at-Lower-Cost-The-Path-to-Continuously-Learning-Health-Care-in-America.aspx Accessed March 25, 2015

Chapter 25

International Health Informatics

ALISON FIELDS

CHRIS PATON

GLEBER NELSON MARQUES

NAOMI MUINGA

STEVE MAGARE

ROBERT HOYT

Learning Objectives

After reading this chapter the reader should be able to:

- Describe innovative international eHealth projects

- Differentiate between different national strategic approaches to health informatics

- Detail the way economic and infrastructure issues impact health informatics projects in low and middle income countries (LMIC)

- Describe how mobile health (mHealth) technology is enabling developing countries to access healthcare information in the absence of formal infrastructure

"Every 10 seconds we lose a child to hunger. This is more than HIV/AIDS, malaria and tuberculosis combined."

Josette Sheeran President and CEO Asia Society

Introduction

It is worth reflecting on the considerable progress the international community has achieved in a relatively short period of time. In regions where data are available, it is clear that the uptake of electronic health records (EHRs) and other health information systems (HISs), has grown considerably in the past 10 years.

In this chapter we will discuss progress and challenges in informatics from multiple developed and developing countries. While that is the stated purpose of the chapter, we should mention that the basic healthcare systems in these countries vary considerably. For example, Australia, Canada, France, Germany, Switzerland and the US typically use the fee-for-service model, whereas the Netherlands, New Zealand, Norway and the UK use a combination of capitation and fee-for-service. Additionally, primary care practices tend to be smaller in size throughout the Netherlands, France, Germany and Switzerland, but much larger in countries such as Australia, Canada, and the United States.[1]

The higher levels of adoption have been realized by smaller, more developed countries that have implemented nationwide eHealth strategies. European countries including Norway, Sweden and Denmark have advanced national health IT (NHIT) systems that have been largely successful. In Australasia, New Zealand is in the process of implementing its second 5-year Health IT Plan.[3] However, some countries have shown that neither considerable financial support nor national strategy guarantees successful implementation of a health informatics program.[4]

Many developed countries such as Norway, the Netherlands, New Zealand, the United Kingdom and Australia are now approaching 100% adoption of EHRs. Table 25.1 displays the EHR adoption rate by primary care physicians in 10 developed or high income countries (HIC), based on a 2012 survey. The table also displays how often online tools are used or are available to patients.[1]

Even with such evidence of great advancement and adoption it is important to recognize that with great strides come growing pains and sometimes great setbacks. For example, some countries such as Denmark have found that rapid adoption comes with its own set of challenges, such as interoperability issues and fragmented patient care.[2] England and Australia suffered strains to eHealth and dismantled their original plans and systems. Both however, are moving forward with new initiatives to serve their populations.

Health Informatics in Europe

Health Informatics in Europe has developed in an irregular fashion with countries usually adopting their own plans and national strategies without a homogenous Pan-European approach. Despite differing frameworks, there has been some uniformity in the development of interoperability standards.[5-6]

European EHR Standards
The ISO EN 13606 standard is defined by the European Committee for Standardization (CEN) that covers all major standards in Europe (not just healthcare standards) and the International Standards Organization (ISO).[7] This standard is designed to achieve "semantic interoperability," meaning different computer systems should not only be able to both read and write data to each other but also understand the meaning of the messages by means of a look-up reference that defines all the different types of data included in the standard.[8]

By using semantic interoperability standards computer systems can more intelligently use data from other systems. For example, a hospital EHR system could use data like blood pressures from a clinic system in its decision support system to issue an alert when inpatient blood pressures values are higher than expected.[8]

Table 25.1: International EHR and Online Tool Adoption

Country	2012 EHR Adoption	Email Access	Online Appointments	Online Refills
Australia	92	20	8	7
Canada	56	11	7	6
France	67	39	17	15
Germany	82	45	22	26
Netherlands	98	46	13	63
New Zealand	97	38	13	25
Norway	98	26	51	53
Switzerland	41	68	30	48
United Kingdom	97	35	40	56
United States	69	34	30	36

In this chapter we will refer to ICT (information and communications technology) as a broad IT umbrella, under which HIT (health information technology) would be included.

Denmark

Denmark is widely held to be a leader in Health Informatics across Europe.[9] It was an early adopter of systems providing patients access to their medical records and their centralized government offered healthcare providers a range of services that assisted the adoption of health IT systems.[10] Part of their success is undoubtedly due to their universal healthcare system that is publicly financed.[11]

Early Adoption and Acceptance
Denmark began their foray into eHealth practices with modest steps beginning in the 1980's with subsidizing the electronic transmission of medical claims in primary care. These early steps encouraged infrastructure development and a foundation for subsequent advancements in HIT.[10] Successive programs were also aided by the 1968 establishment of Denmark's central citizen registry. This had allowed citizens to become incrementally comfortable with their personal information being held privately and securely in a high quality and readily accessible electronic manner.[12]

Medcom
MedCom is a not-for-profit publicly financed company owned by the Danish government that provides the infrastructure to allow the secure transfer of digital messages between all patients, hospitals, labs and pharmacies, such as, "discharge letters, referrals, lab test orders, e-prescriptions and insurance reimbursement." The system has been very successful with most healthcare documents now being transmitted in digital format.[14-15]

Interoperability
While national standards provide the framework for all EHR systems, there are 5 regions that each subscribe to their own EHR systems but each one is interoperable on a national level. By 2017 there will be 2 regions that will be using EPIC EHR, further reducing the number of EHR systems down to 4 for the whole country.[16]

Sundhed.dk
Sundhed.dk is the official health portal for Denmark that allows patients to access their health data, make appointments and access informational resources, such as, disease guides. They can also see waiting list times, ratings of public hospitals, and access online patient support groups. Doctors and other healthcare professionals can also access patient data through the "e-Journalen" and "Shared Medication Record." These tools essentially allow access to health data that was previously more difficult to access.[13] Figure 25.1 displays how Sundhed.dk is organized.

Figure 25.1: Sundhed.dk organization

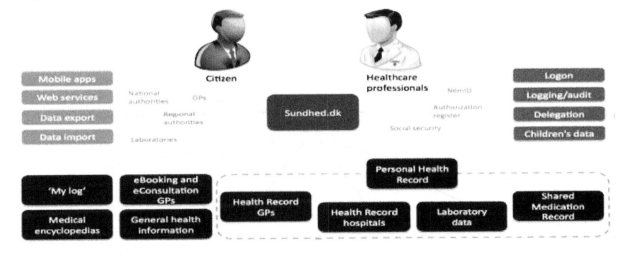

Healthy mHealth Sector

Denmark has conditions that has made mobile health (mHealth) thrive, including at least 70% smartphone use with more than 90% considering themselves regular users.[17] Denmark has been listed as the first choice for mHealth business based on readiness, reputation, practitioner acceptance, and level of digitization.[18] In September 2015, the Danish Digitization Authority reported that it would invest $3.3 million in 5 new telehealth projects to address home physiotherapy to reduce readmissions, home heart monitoring, a virtual endocrinology outpatient clinic, acute care monitoring for elders, and communication between type 1 diabetes patients and healthcare staff.[19] An active and engaging health tech environment has allowed such innovations as Monsenso's app for the treatment and analysis of mental health disorders and "Virtual Rehabilitation" for physiotherapy.[20-21]

Strategy for Success

Factors that have contributed to globally recognized success in Denmark include universal healthcare, regional record systems, adhering to national standards of interoperability, unique patient identifiers, small population size, higher incomes, and high population access to the Internet.[2,22]

United Kingdom (UK)

HIT Strategy

Like Denmark, HIT has been guided by a National Health Service (NHS). Patients are assigned unique identifiers, and electronic prescriptions are widely used. However, this appears to be where the similarities end. In 1998 the NHS initially planned to provide EHRs, online access to health services, and a national system to develop and coordinate HIT services.[23] However, subsequent implementation has been a costly strain.

The National Programme for IT

Perhaps the best-known health IT project in Europe is UK's National Programme for IT (NPfIT). This was an ambitious program initiated by former Prime Minister Tony Blair to create a centrally managed system of IT to cover the whole of the NHS in England.[24] At the core of the project was the development of the "Spine", a central database that would hold the records of some 50 million patients.[25] The project commissioned a number of large IT companies to provide EHR and information systems to hospitals across the country.

The program has been largely viewed as a failure, with over $12.7 billion pounds spent over the course of 9 years with little software or infrastructure delivered, while functioning only at "base functionality."[24,26] Multiple impediments have been cited, including haste, architecture issues, and lack of motivation, leadership and technical expertise.[24,27-28]

NHS Identifier

Unique patient identifiers are seen as important in enhancing patient safety and preventing adverse events.[29] In the UK, the NHS number is a unique patient identifier used to access health services and to match individuals to their clinical records. With accuracy and completeness of this identifier ranging from 90%-99% across all levels of care it is routinely used to correctly identify patients. From a research perspective, the NHS number is of particular relevance as it is often used to link clinical data to other datasets, such as socioeconomic or mortality data.[30] Unlike many other countries, the NHS number is consistent and highly unlikely to change over the lifetime of an individual.[31]

However, there is a growing need in the UK to assure patient privacy and confidentiality, thus identifiable data is pseudonymised, or replaced with codes that are only recognizable to treating physicians and not researchers. The UK Data Protection Act of 1998 limits the sharing of confidential data to only those instances where explicit consent has been provided and is rigorously applied across the UK to prevent unethical use of identifiable data through legislation. These restrictions while aimed at maintaining confidentiality and privacy of personal information, have also been criticized for causing many research endeavors to be carried out less efficiently or not at all.[32]

TeleHealth in the UK

The UK aspires to increase the use of telecommunication technologies to monitor patients at a distance.[33] This has been seen as a cost effective alternative to long-term institutional care. Providing care in the home with an intention to support self-management of patient care requirements is one-way telehealth. Driven by the Department of Health, the UK undertook a clinical trial (Whole Systems Demonstrator Programme) which was possibly the largest telehealth and telecare trial in the world aimed at testing the efficacy of telehealth use.[34] In an early analysis of results on chronic conditions, such as diabetes and heart failure, the study showed an overall improvement in health measures and a significant reduction in costs, emergency visits, hospital admissions and mortality rates.

However, other recent studies have showed no significant effect on patient outcomes with the use of home monitoring, as compared to usual care.[35] This initiative is also discussed in chapter 18 on TeleHealth in the textbook. A majority of studies of teleHealth interventions did not adequately define nor address the meaning or differences in "usual care" and trials failed to take into account the new innovations in technology such as smartphones, tablets and the Internet.

Healthspace

Healthspace was envisioned in 2006 to be a patient health record web portal that would allow patients to access and edit their own records, create their own health data, provide a platform for secure communication to providers, and make appointments.[36] The program suffered from low strategic importance, limited functionality and as described by Dr. Charles Gutteridge, the National Director for Informatics at the Ministry of Health, " just too difficult"."[37] By December 2012 Healthspace was officially shut down as a result of low interest, as less than 0.01% signed up for comprehensive accounts.[38]

Post NPfIT

In 2011 the UK government announced that NPfIT would be dismantled in favor of a system of devolved procurement to hospitals, provided that local programs adhere to "nationally specified technical and professional standards."[39] As of mid-2016, practices and hospitals are encouraged to utilize electronic records and forms, but there is no requirement to do so, and medical records (from hospitals, general practitioners and specialists) are not integrated.[40]

Despite a lack of a requirement for a specific EHR or definitive interoperability standards, 96% of physicians use some sort of electronic patient record in their practice; 38% can share patient summaries and test results with physicians outside of their practice and 85%

receive electronic notification of improper drug dosage or interaction.[40]

In October 2014, the NHS released the Five Year Forward View Plan outlining the future of their health care system. Included in the brief document was the creation of a National Information Board to increase transparency, expand health apps, advance the number of interoperable EHRs, and bolster the use of technologies, such as smartphones by staff and patients.[41]

The NHS published "Personalized Health and Care 2020" in 2014 and it built upon the Five Year Forward Plan. They referred to it as a "Framework for Action," rather than a formalized plan or strategy.[39] Follow up "roadmaps" were then published in June 2015 outlining achievements and regional conferences were scheduled throughout July 2015 for patients and practitioners to provide feedback on the direction of the roadmap, barriers to delivery, and incentives provided to achieve success.[42]

The Interoperability Handbook was published by the NHS in September 2015 as a resource to aid interoperability, standards and implementation.[43]

To add to the challenges experienced by the UK, the Cambridge University System implemented Epic EHR, but experienced a myriad of difficulties.[44]

Positives in the mHealth market

Despite difficulties in the overall eHealth design and implementation, the UK has received recognition as an excellent country for mHealth.[18] Examples of UK's mHealth programs include: iPlato's Patient Care Messaging service; mHealth Assist that provides a platform for communication, information, and support for those with chronic conditions; and Personaltechmd's exploration of a new wearable to aid dementia patients, their families, and practitioners.[45-47]

Germany

Early Adopter of Medical Informatics

Germany was once a leader in advancing Health Informatics. In 1949 Gustav Wagner formed the world's first professional organization for Medical Informatics in Germany.[48] Since then,

Germany has gone on to build a mature eHealth system with 92% electronic records and a strong medical technology sector. Despite their strengths, Germany still struggles to improve in the areas of health information exchange, percentage of General Practitioners (GPs) with websites and e-prescribing.[49]

German Smart Card eGK
A critical component of Germany's eHealth infrastructure is the national Electronic Health Card (Elektronische Gesundheitskarte OR eGK) (Figure 25.2).[50] The original card was designed for 72 million customers, doctors, hospitals, and pharmacies. It was intended to hold insurance and prescription information and optionally house data on drug intake, chronic diseases, blood type, operations, lab results, and a disease diary.[51]

Figure 25.2. German Smart Card

Gesundheitskarte

G 1

AOK

Erika Mustermann
106415300   A123456789
Versicherung   Versichertennummer

However, the massive and expensive (1.7 billion EUR) roll out of the universal e-card was virtually suspended early in 2010 due to difficulties surrounding the complex nature of the e-card plan, data privacy concerns, and opposition from providers.[52]

The eGK was relaunched in October 2011 to six German states and the general population in 2013.[53] The initially ambitious aim of including medical data and e-prescribing information has been scaled back, but the card is designed to support future national eHealth projects.[54] The card facilitates identification and access to services, such as, an electronic patient file, diagnoses, labs and x-rays, and e-prescribing. There is also an added layer of security that requires a patient to enter their PIN number in order for the e-card to operate. Since the introduction of the updated eGK 95% of the population has been equipped with the new cards.[55]

Unleashing Interoperability
In 2014 Germany released the "Digital Agenda 2014-2017" underscoring the need for increased interoperability, as well as eHealth innovations. To that end, they announced the implementation of a standards framework at the federal level to increase interoperability.[56] As part of the restructuring the new "eHealth Initiative" was announced as an independent working party to address the needs defined in the Digital Agenda. The two cornerstones are a study to determine concrete ways to remove interoperability barriers between their 200 different HIT systems and the creation of a national telemedicine portal.[57] The original eHealth infrastructure leaned heavily on the eGK, however the Digital Agenda, the eHealth Initiative, and the eHealth Act all aim to establish a singular telematics infrastructure to function in conjunction with the electronic health card.[58]

In addition to announcing these changes, the draft bill also introduces incentives for physicians and hospitals to cooperate with new innovations, such as, creating emergency data or electronic discharge letters.[59]

mHealth App uncertainty
The mHealth market for apps in Germany is mixed and considered the most controversial of the EU countries.[18] There have been pockets of positive news for apps, such as, Caterna Vision Therapy for amblyopia treatment, the first reimbursable app.[60]

France

Early Groundwork
As early as 1978, the French National Social Security System planned an evolutionary leap to use smart cards and electronic care sheets. It took 20 years for an initial roll out of the Sesame Vitale card but during that time a strong backbone of optical fiber was installed.[61]

DMP
The Dossier Médical Personnel (DMP) is a nationwide online patient health information project that was initially launched in 2004, halted in 2006, and re-launched in 2011.[62] The record includes: patient history, allergies, medication history and lab results. Records are accessible on the Internet and patients can choose which providers can view their records and allow full access or read- only status.[63] It

was designed to aid in the coordination of care between providers, reduce duplication of actions and documentation and prevent drug and treatment interactions. Providers are also required to use office EHR software that is interoperable with DMP.[64]

Similar to HealthSpace, the DMP project is voluntary (opt-in) and has not garnered much use so far. As of December 2013 less than 1% of the population had a DMP file and the cost thus far is reportedly in excess of $210 million EUR. However, unlike HealthSpace, the project continues to move forward.[38]

Carte Vitale Card

Carte (Sesame) Vitale is also a staple in the French eHealth system. The original card was introduced in 1998, strictly for insurance billing purposes, but was envisioned for varied purposes. However, after several generations, the Vitale 2 cards continue to function primarily as a paperless billing and reimbursement vehicle. It is important to note that its use has been linked to reduced administrative costs which in turn have reduced overall treatment costs.[65]

The CPS Smartcard

The CPS card is France's healthcare professional card that contains data on the provider's identity, profession, specialty, and hospital or facility affiliation. They afford a provider the ability to sign and send electronic forms, add and edit EHRs, and access a secure messaging platform.[66]

Potential mHealth growth

Like Germany, France is considered a complicated market for mHealth. Low adoption has plagued mHealth in France, however the market is seen as poised for large potential growth in the future.[18]

Health Informatics Systems Interoperability Framework (HIS-IF)

HIS-IF adapts the international Integrating the Healthcare Enterprise (IHE) profiles to the French context. The main purpose of implementing the HIS-IF across the country is to provide hospitals and clinics an avenue to communicate with the DMP system and to enable the secure creation and storage of personal health records (PHRs). It is a central framework that addresses both technical and semantic interoperability; thus, allowing vendors to concentrate on details and specialty functions while removing the guesswork for providers wanting to purchase compatible products.[67]

Health Informatics in Australasia

Australia and New Zealand (NZ) have both undergone significant developments in Health IT over the past decade. The focus in NZ has been on the development and implementation of a NHIT Plan. In Australia the National eHealth Transition Authority (NeHTA) has implemented a number of projects, the highest profile of which was the Personally Controlled Electronic Health Record (PCEHR) that later evolved into My Health Record.

New Zealand

NZ has been a well-recognized provider of eHealth services since the 1980s when regional hospitals began integrating electronic administrative systems. Their initial steps towards eHealth began in 1992 with the creation of unique patient identifiers and an interoperability framework.[68]

NZ has a top-down approach to standardization and interoperability and the government is directly involved with the development of HIT, runs the Health Information Standards Organization (HISO), and updates guidelines regularly. The NHIT Plan aims to implement common health IT platforms across each region of NZ for managing patient administration and clinical information. Each of their 4 regions are able to choose their own EHR, repository, and support systems but are required to integrate with the National Health Index systems.[69]

Currently, their use of electronic means for provider follow-up and preventative care reminders is 92% and physician use of EHR is nearly 100%, while also ranking high in areas of patient centered care and coordinated care. Patient portals are also accessible to view medical records, set appointments, and renew prescriptions.[11]

There are multiple different add-on platforms that the country has introduced in recent years with great success. One such program is the national child IT platform to track important health milestones for children.[70]

Orion Health

Orion Health is an NZ based eHealth technology company founded in 1993 that has become a global force in providing EHRs and other healthcare solutions. They provide services to the majority of health districts in NZ and their products have been implemented in over 30 countries.[71]

Interoperability

While their use of eHealth in individual settings is optimal, data sharing between facilities and specialties is limited.[11] Concerned by the lack of functionality on a national scale, the Minister of Health in 2015 announced a plan to move to a single countrywide EHR system with portal and mobile capabilities. In April 2016, the Ministry released the *New Zealand Health Strategy 2016* with one of its five goals being a "smart system." They specifically plan to support a universal EHR, patient portals, analytics and a health app formulary.[72]

mHealth Innovation

NZ is reportedly leading the way with mHealth. The University of Auckland conducted the first study of text messages for: smoking cessation, advice for pregnant women and families, diabetes patient self-management, and a youth-line for the text friendly teen population to help with issues such as bullying, relationships and sex, drugs, and abuse and violence.[73] Another interesting app is "Beating the Blues," a mental health treatment program for depression and anxiety that is available through a patient's GP.[74]

Australia

Australia's eHealth plans have been slowly developing since 1993 when they created the National Health Information Agreement (NHIA) "to develop, collect, and exchange uniform health data, information, and analysis tools." From 1993 until 2007 they laid the foundation for nationwide eHealth in 1999 with their Health Information Action Plan, in 2005 by forming the National E-Health Transition Authority (NEHTA), and in 2007 with legislative changes to update codes and guidelines.[75]

In 2010 AU passed the Healthcare Identifiers Act to assign identifiers to individuals, providers, and organizations.[76] The PCEHR was established in 2012 by the NEHTA; designed to used interoperability standards such as HL7 and the Cross Enterprise Document Sharing (XDS)

profile from Integrating the Health Enterprise (IHE) to allow medical professionals to share a variety of clinical information with patients and each other.[77]

However, their progress has not been without growing pains. PCEHR was plagued with low participation, high cost, and eventually replacement. Some practitioners felt that the program had been hijacked and bogged down by lawyers and legislators, did not serve the needs of clinicians or patients, lacked government support, and suffered multiple access and functionality issues.[78] PCEHR was similar to the DMP in France in that it had a significantly lower number of participants along with a reportedly hefty price tag.

My Health Record system

In 2015 it was announced that the PCEHR would undergo an overhaul. It was modified to become My Health Record, an online PHR. It has the ability to store medication summaries, claims history, hospital discharge summaries, imaging reports, organ donation statuses and pathology reports. As of May 2016, 2.7 million patients and 8500 providers had joined the service. The physician incentive program will be tied to participation in the new record system. It is too early to know the impact of this system on patients and clinicians.[79]

mHealth in the AU

mHealth is in its infancy but has potential, as there are an estimated 15 million Australians (~63%) with smartphones.[80] mHealth technology is growing and thriving within AU and is seeing successful exportation as well. Telstra's ReadyCare is a teleHealth service launched in 2015, offering connection to a GP through a 1-800 number or an app. Patients will be able to discuss their issues, upload photos, and receive treatment including prescriptions.[81] Another mHealth system is KinetiGraph for Parkinson's disease patients that is worn like a wristwatch and logs activity and reminds patients when to take their medications.[82]

Health Informatics in Africa

There are a wide range of health informatics initiatives underway in Africa that are quite different from initiatives in Europe and the Americas. Due to a lack of IT infrastructure and financial resources, many projects make use of open source software and mobile technology (mHealth).

Prioritizing eHealth in Africa

Historically, Africa's health agenda was dominated by donor supported programs that mainly focused on a narrow and high profile disease specific areas, such as HIV, TB, or malaria.[83] The interventions frequently used standalone HIT to support their programs and to track individuals, monitor the useful indicators, and to store the health information they collected. The interventions were often ad hoc, in response to epidemics, had a short to medium term outlook, and subsisted on inadequate ownership from beneficiary governments. While the investments had impact and uplifted the health status of populations across many African countries, there was poor coordination, which in-turn resulted in a replication of efforts and disjointed approaches to eHealth.[84-86]

A significant advantage realized was that resources that would have otherwise been spent on tackling crucial healthcare issues, were now funded by international bodies, thus making these resources available to African governments for alternative investment. Investments in HISs fell lower on the list of priorities as policy makers leaned towards other national issues that would maximize the general populations' state of wellbeing, rather than investing in eHealth which was not seen as a direct and immediate need.

In those rare cases where these funds were invested in eHealth, most initiatives rarely went past the pilot phase and were not well implemented or documented. This made it difficult for countries to learn from the experiences of other similar countries.[87] Furthermore, most programs lacked a comprehensive evaluation mechanism. These limitations led to some failures, and a lag in eHealth adoption, compared to developed countries.

Current state of eHealth

Over the past 5-10 years, many African countries have made significant strides and rolled out more coordinated health information systems. National strategies to standardize eHealth have now been developed across Africa. These efforts can be seen in multiple African countries. With governments and relevant institutions such as Ministries of Health playing more pivotal roles, the progress achieved is beginning to be felt.[88]

This has not been without its challenges and failures in eHealth implementations are more heavily felt in resource constrained environments. In the past couple of years, significant progress has been made towards standardizing developments in eHealth. Ad-hoc project specific approaches are being gradually refined and initiatives are taking a more systematic approach by establishing government backed eHealth departments, dedicated IT infrastructure and funding.

Less expensive and more sustainable open source software (OSS) systems, such as OpenMRS and other health applications have been adopted widely. These technologies are being used extensively in public facilities and remote communities to record health events and connect rural populations to skilled health workers. Local customization of such widely available open source solutions and the proliferation of mobile phones minimizes the financial outlay and are proving to be a preferred and more sustainable strategy over the long-term.

OpenMRS

The project is backed by I-TECH at the University of Washington and was co-founded by the Regenstrief Institute at the University of Indiana and a number of partner organizations across Africa. OpenMRS is now being developed into a full EHR solution that can be used by both hospitals and clinics in Africa, although it is primarily still used for smaller clinics. The most recent features include patient summaries, vital sign capture, outpatient or inpatient capture, and the ability to add diagnoses using a coded or non-coded methodology (Figure 25.3).[92]

Figure 25.3 OpenMRS

DHIS: District Health Information System2

One of the most significant health informatics successes in Africa in recent years has been the adoption of the open source District Health Information System (DHIS).[89] This cloud-based service allows countries such as Kenya, Tanzania, Uganda, Rwanda, Ghana, and Liberia to manage complex data collected from healthcare facilities across their respective countries (Figure 25.34).

DHIS2 is the current version of the open source system which allows anyone to download, install and change the software for free. The project is coordinated with the University of Oslo in Norway and the Health Information Systems Programme (HISP).[90] DHIS2 pulls in data from local hospitals, either by manually entering statistics about the functions of the hospitals (how many patients, what kind of treatments, etc.) or by integrating DHIS2 with an EHR system, such as OpenMRS. Once in the system, users are able to use DHIS2 to explore the data, run reports, and generate visualizations.

Figure 25.4. DHIS2 Adoption

mHealth in Africa

African countries have experienced a rapid adoption of mobile technologies in recent years. The introduction of mobile phones with SMS capabilities has enabled a large number of highly successful initiatives, from reminding patients when to attend HIV clinics to sending out regular advice for expectant mothers.

As 3G network connectivity continues to spread and tablet PCs and smartphones increase in adoption, new opportunities are emerging in developing countries for a wireless Internet infrastructure that has the potential to "leapfrog" the huge investment in fiber-optic and copper networks that have been installed in the developed world over the last few decades. With the opportunity for 4G and 5G networks just around the corner, there is a significant opportunity to develop bandwidth intensive healthcare solutions (EHR, imaging, and video-conferencing) for a relatively low cost compared to the vast sums invested in hard-wired networks.

In addition, the introduction of low cost satellite Internet connections such as those offered by Facebook's Internet.org, SpaceX, and Project Loon offer intriguing new ways to accelerate connectivity in developing nations and may be able to bridge the gap in connectivity in areas where 3G networks are not currently available.[93-95]

Mobile Phone Subscriptions

The International Telecommunications Union (ITU) has collated statistics on comparative mobile phone adoption in developed and developing countries. (Figure 25.5)

Figure 25.5 Global ICT use

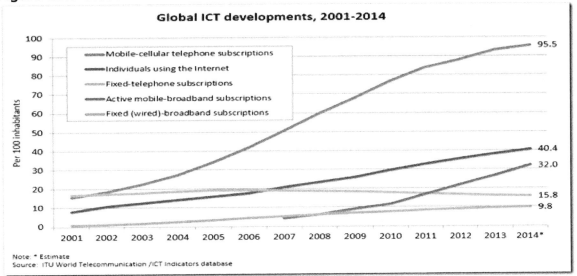

The proportion of users in the developing world is accelerating quickly, rising from just 35% of subscriptions in the year 2000, to 78% of subscriptions in 2014.

The number of people using their phones and tablet PCs to access the Internet is also increasing more rapidly in developing countries than the developed world rising from 20% in 2008 to 55% in 2014.[96]

Current mHealth Options

Medic Mobile
Medic Mobile is a non-profit organization based in San Francisco that operates in 15 sub-Saharan African countries. It developed a platform for delivering SMS (text) messages based on the open source FrontlineSMS software that was originally developed by conservation charities in Africa. Medic Mobile uses the software, combined with a suite of EHR modules, to give healthcare workers the ability to use SMS to help with vaccination efforts, keep track of patients, send out appointment reminders, and conduct research.[97-98]

Vodafone and mHealth Alliance
Vodafone and United States Agency for International Development (USAID) have been working with partner organizations on a number of mHealth projects in low resource settings through the mHealth Alliance initiative. In

Africa, the Alliance has been involved in a number of initiatives:

- **GAVI The Vaccine Alliance** is a 3-year partnership between Vodafone and health ministries across sub-Saharan Africa to use mobile phones to improve rates of immunization. Proposed methods include "alerting mothers to the availability of vaccinations by text message, enabling health workers to access health records and schedule appointments through their phones, and helping health facilities in remote locations monitor (vaccine) stocks."[99-101]

- **Vaccines in Mozambique.** Vodafone has been working with the global healthcare provider GSK, Save the Children, and the Mozambique Ministry of Health to implement a project using mobile technology to raise awareness of vaccination among expectant mothers.[102]

- **SMS for Life.** Vodafone, Novartis and other partners have developed an mHealth initiative in Tanzania with Roll Back Malaria. [103] The SMS for Life project aims to use mobile SMS messages to help healthcare workers keep track of malaria drug stock levels across three districts in Tanzania. This initiative has achieved a high reporting compliance rate in Tanzania and Kenya.[104]

- **Comprehensive Community Based Rehabilitation in Tanzania (CCBRT)** is a partnership between the Vodafone Foundation and m-Pesa, the mobile money transfer and microfinancing service, to address the problem of obstetric fistula that causes maternal incontinence post-childbirth in many African women. The CCBRT hospital uses the m-Pesa system to enable mothers to fund their travel to hospitals for surgery to correct the condition. The project aims to enable 3,000 women a year to get access to treatment.[105-106]

Mobile Alliance for Maternal Action

Mobile Alliance for Maternal Action (MAMA) is a successful mHealth project that has been working with a wide range of partner organizations, including the United National Foundation, USAID, Johnson and Johnson, the mHealth Alliance, and BabyCenter. MAMA aims to improve healthcare provisions for new and expectant mothers in low-resource settings through the use of SMS messages containing health advice for them and their newborn children.[107]

As can be seen by the range of projects previously described, many different types of organizations are using relatively low-tech solutions to improve healthcare provision in Africa. As the technology improves, it will be interesting to see how new features of smartphones such as high resolution cameras, fingerprint identification systems, and faster and more powerful applications will be used in future projects.

According to Grices's conversational maxims, the nature of SMS messaging other than for the very simplest of instruction may be open to multiple interpretations.[104,108] Because text messaging is not privy to essential nonverbal cues, they may lack indications of urgency or importance. The way a message is framed can also affect whether a person is receptive to making a behavior change or not.[109] It is crucial that text messages are written in the most appropriate way for the population, including ethical and cultural considerations. Despite challenges in establishing text message content, a recent study showed simple text reminders increased women's attendance of breast cancer screenings, as well as in adherence to taking HIV/ART medication.[84, 110]

However, overall reviews performed for interventions applied in more developed countries, such as in the use of mobile text messaging in self-management of long term illnesses, reminders to attend clinical appointments, and messages to communicate results of medical investigations, have showed only a moderate effect on outcomes or behavior.[111-113] Existing literature has proven the efficacy of text messaging for appointments or drug reminders in adult populations.[111, 114-115]

Kenya

Kenya's eHealth strategy was developed through a consultative process and is anchored by Vision 2030.[116-117] The strategy promotes the delivery of efficient health services enabled by ICT and the implementation of the official standards and guidelines which were developed to harmonize the various donor-funded implementations. The country continues to embrace open source solutions with use of OpenMRS and DHIS2, described earlier. A 2011 study showed that proprietary and open source systems are used in almost equal measure. Proprietary systems are mainly used in private hospitals and faith-based organizations which mainly purchase off the shelf proprietary solutions to manage their facilities.[118]

Kenya, like many LMIC, has "leapfrogged" most traditional eHealth activities. New mobile or cloud systems have been adopted to replace paper systems without investment in PC or mainframe systems, routinely adopted by HIC. Wireless technologies have been shown to reduce the administrative burden on health care workers and resulted in improved patient care.[119] Web and telephone based health consultations, such as an e-consultation service *Sema Doc* launched by a leading mobile phone company, is also gaining traction in urban and semi-urban areas.[120-121]

Systematic EHR use outside of donor funded health programs remains lacking and where they are used, the systems tend to be immature. Kenya continues to face limitations in eHealth and ICT financing, medical personnel shortages, regulation, and legislation to protect privacy and security of patient data.[122]

All of these factors have considerable impact on healthcare and by extension, eHealth. The Kenyan government's national strategy

"Towards a digital Kenya" aims by the year 2017, to make broadband available to all health centers country wide.[123] This will assist greatly in positioning local/rural facilities to realize the potential of eHealth and benefit from teleconsultations through telemedicine.

South Africa

Similar to Kenya, the South African health information system was fragmented and faced challenges of interoperability where automated systems existed. This then led to the development of an eHealth Strategy to be used as a guide in integrating HIT for health delivery. South Africa through the National Department of Health (NDoH) commissioned the development of a NEMRS.[124]

Rwanda

Rwanda is one of the poorest, smallest, and most densely populated countries in Africa.[125-126] Nevertheless, it has made significant progress in deploying a NHIS driven by a strategic plan for eHealth, championed by the Ministry of Health and Sanitation. The strategy stresses the need for governance in the development of eHealth infrastructure to aid in the efforts against proliferation of fragmented and piecemeal HISs, placing system interoperability high on the list of priorities.[127-128] It is based on an eHealth Enterprise Architecture Framework which functions as a broad roadmap to ensure interoperability between all databases within the health sector. Rwanda is yet to meet any of its Millennium Development Goals (MDG) targets but it has, over the last decade, halved the maternal mortality rate and made significant progress in all the other MDG indicators. It is believed to be on track to achieve most of the MDGs by the end of the decade.[127,129-130] The Government's vision to further utilize technology in healthcare, has also made Rwanda a pioneer in integrating technology into the healthcare system, both regionally and also in Africa.[131] Regional health experts and officials from the East African Community (EAC) have recognized this progress and made it the lead for eHealth and technology use in healthcare.[132] Rwanda has also adopted DHIS2 and OpenMRS which are seen as more established and stable open source web based platforms. With the OpenMRS platform being piloted successfully in 24 health facilities in 2013, it is being extended nationwide.

Various mHealth initiatives are also important components of this strategy which feed into Rwanda's Health Management Information System. Maternal and newborn health, HIV, and malaria epidemics have been a central target for Rwanda's HIS efforts. The focus is now gradually shifting from infectious/communicable diseases to non-communicable and chronic conditions. Community health workers (CHWs), numbering over 45,000 operating at the village level, provide first line health care delivery.[130] CHWs increase efficiency in reporting disease incidences and delivering care and advice to the populace. To do this they have embraced low cost but effective mHealth solutions such as:

RapidSMS
RapidSMS is an interactive two-way short messaging service (SMS) based information system developed to support the documentation of pregnancies and related events including emergencies within the community. RapidSMS has proved effective in preventing maternal deaths with CHWs being central in collecting maternal indicators and recording births for purposes of aggregation and central storage.[133-134]

TRACnet
Treatment and Research AIDS Centre and other extensions, such as TRACnet Plus, accessed largely through the mobile phone and also through the Internet, have been used to capture health events.[135] These systems enable practitioners in remote HIV/AIDS clinics to communicate with administrative units to manage patient information and submit reports. The systems have enabled real time access to critical tracking indicators, such as HIV transmission patterns, drug stock levels, etc. and have reduced the time for result reporting.[136] TRACnet is currently being extended beyond HIV to capture over 23 other indicators for communicable diseases. This includes monthly monitoring of infectious diseases including TB and malaria. TRACnet SMS messages also feed centrally into the electronic system for integrated disease surveillance in Rwanda.[137] Criticisms of the technology include inconsistencies in recording of clinical data and lack of recent studies needed to gauge the accuracy and quality of the data collected.[138]

SIScom
This program, accessed primarily by mobile phone, allows CHWs involved with maternal

health to send patient data via text messages and report on any incidents during the pregnancy to the district hospitals. This system has enabled better patient/maternal follow-up, improved mother and child health, and reduced mortality rates significantly.[139]

Health Informatics in Asia

China

History
China holds the distinction of having not only the world's largest population but also the largest elderly population.[140] These factors have put pressure on the government, healthcare organizations, and providers to adopt some form of digitized health care. Computerization of administrative and computer data in hospitals began in the late 1980's with single computer usage.[141] During the 1990s China began increasing IT development in multiple industries through several "Golden Projects." In 1995 China's Ministry of Public Health announced the "Golden Health Project" aimed specifically at healthcare to increase connectivity of hospitals, medical research, and medical education.[142-143]

The Chinese government announced another national health reform plan in 2009 to improve medical coverage to 90% of the population, increase access to essential drugs at lower prices through providers outside of hospitals, and increase use of local healthcare centers as primary contact points rather than larger central hospitals. Increased adoption and interoperability of HIT is an integral part of addressing all the other aspects of their plan.[144] In 2012, the MOH released the 12th five-year plan that included the establishment of a NHIS.[145]

EHR Adoption
As part of the 2009 reform, a nationwide EHR was announced with coverage goals of 30% rural and 50% urban by 2011.[146]As of 2014, China was still short of their original goals with 20% rural and 30% urban adoption rates. However, they are planning continued expansion of EHR to 50% rural and 50% urban by 2020.[145]

Interoperability
While most healthcare providers have some form of EHR, a significant number of them are not interoperable and there is still no unified national system in place to integrate patient records. While hospitals are linked to insurance for payment administration, they are frequently not linked to one another, even in the same region and when owned by the same entity. Currently, patients do not use EHRs or online services for patient access, appointment booking, prescription refills, and messaging.[147]

mHealth
Mobile phone ownership in China tops 1.17 billion devices and subscription levels are high at 92% of the population, incorporating even rural portions of China. The role of mHealth is expanding into primary care as a means to bridge the gap in rural and underserved areas. Providers are currently allowed to offer electronic consultation, without treatment or prescribing medication.[148]

China is home to nearly 100 million diabetic patients leading to the development and ample funding of mHealth apps and plugins to monitor and manage glucose levels, dietary intake, and medications including Weitang and Dnurse.[149-150] Other mHealth app sectors looking to take advantage of the market potential are electronic consultations, intelligent mobile medical devices, and rural child immunization.[151] To address the growing need for mHealth that specifically addresses the nuances necessary for the Chinese market, the Stanford Center at Peking University launched "Digital Health Bootcamps" in 2015. These events bring together hospitals, programmers, designers, marketers, and business experts to develop mHealth apps.[152]

Patients are responding to the convenience and ease of mHealth technologies with a rise in use, from 66 million people in 2014 to 138 million in 2015.[153] High mobile use, rising need, and increased funding will ensure that growth will continue for apps, wearables, and monitoring devices.[148]

Telemedicine
China is now exploring telemedicine solutions to address a quality and distribution gap between urban and rural areas.[154] With the continued increase in their already high population of elderly, telemedicine is also seen as a way to care for their aging patients while reducing pressure at overburdened urban hospitals, already hindered by provider shortages.[155] Several companies have stepped up to offer online platforms including Haodiafu, Chunyu Doctor, and DXY while doctors are ramping up their use

of online platforms to communicate with peers and patients.[156]

Cultural complexity and Social makeup

The Chinese culture is infused with a reverence for tradition, conservatism, and caution. While these values may hold their community together, they may also be impeding their progress towards innovation and acceptance of new health technologies.[157] In addition, the large rural makeup of the population and providers has been a barrier to adoption and use of EHR, even when available. Lack of understanding, comfort with electronics, and security concerns has prevented wider adoption of HIT.[146]

Traditional Chinese Medicine (TCM)

Healthcare in China is a chimera with a blend of TCM practices coexisting with Western medicine. While there has been some embrace of Western medicine, use of TCM hospitals and clinics is on the rise, accounting for 17.9% of all visits (530 million) and one third of the total medical industry financially.[158] Hsu et al. noted barriers to HIT implementation by TCM practitioners, including lack of adoption and limited computer experience, concerns surrounding patient confidentiality, trepidation over potential interference with practice workflow, and reluctance to share propriety techniques and formulas.[159] Additionally, current HIT designed for Western medicine does not appear to be "plug and play" with TCM. The use and interface of the two may be similar; however, "TCM EMRs required a different logic and very different terminology."[160]

Limited research

One of the obstacles potentially reducing progress is a lack of research into health technology assessment and usability. Research in these areas ensures accurate policy, funding, creation, and application of new health technologies.[161] There is a shortage of staff to conduct research and an even more limited number of those are clinically well trained.[162] In addition, there has not been a streamlined or coordinated effort to assess and monitor the past or future application of HIT.[161]

Market Complexity

Translational issues, both language and cultural, have been cited by large companies such as Epic, Accenture, and KLAS Enterprises as additional impediments to the expansion of HIT in China.[160] The U.K. Digital Health Demonstration Centre is also trying to take some of the guesswork out of importing HIT services into China, demonstrating available U.K. services to government, hospital, healthcare, and private organizations, and to foster collaboration between the countries.[162]

Japan

History

Japan began building the base infrastructure of digital health in the 1970's in an effort to address billing administration and management.[163] The 1980s saw the formation of the Japan Association for Medical Informatics (JAMI).[164] In the 1990s Health Information Exchanges (HIEs) were formed to focus on data sharing between hospitals, clinics, and laboratories, but not regions.[163] The 2000s ushered in Electronic Medical Records (EMRs) as a part of their "e-Japan Strategy."[165]

Current Systems

One exemplary healthcare system in the Nagasaki Prefecture is AjisaiNet. It is Japan's largest health care network linking 27 hospitals and 434 clinics, pharmacies, and labs for over 38,000 patients.[166] Each provider determines level of sharing such as providing access to lab images while privatizing physician notes. It is considered a vanguard system that may pave the way to greater regional and eventually national coverage.[166-168]

Japan's "My Hospital Everywhere" was launched in 2013 to allow patients and providers the ability to store and access EHRs across the country.[168] In spite of EHR initiatives, as of 2011/2012 large 400+ bed hospitals had only a 50% EMR adoption rate, while small 20-199 bed clinics only reported 13% EHR adoption.[169] The lack of uniformly and nationally available health records was readily apparent after the 2011 earthquake off the coast of Japan and resulting tsunamis. Hospital facilities were destroyed, paper records were unavailable, and providers and patients alike were in an upheaval.[170]

Unique Identifiers

Beginning in 2016 a new citizen identification numbering system will be rolled out in Japan. While not the initial and primary purpose of the system, medical record creation and access is planned for launch in 2018.[171]

mHhealth

Japan is seemingly slow in adapting and adopting mHealth measures.[172] The current bulk of mHealth in Japan is providing connectivity between seniors in remote rural areas sending health data such as weight, activity levels, and blood pressure to their providers.[173] One such movement is the joint venture between the Japan Post and Apple to provide the elderly with iPads enhanced with a variety of health apps to address senior specific topics.[174]

Telemedicine
In Japan telemedicine services has mostly been limited to chronically ill patients in very rural areas. A major obstacle is health insurance guidelines that restrict practitioner's earnings to a fraction of their normal fee when providing telemedicine services.[175]

Issues
One major hindrance to the growth of eHealth in Japan has been the stringent regulations placed on providers that prevent discussing specific matters of health or disease via phone or email.[176] Additionally, lack of continuous funding appears to have been a limiting factor in the spread of breadth and depth of HIT.[163] Despite the fact that growth and adoption of digital health has been limited in Japan, there has been a movement for health care funding and management to shift from public to private to pursue profit and efficiency.

India

Healthcare in India is a paradox. They have well trained physicians and can offer medical tourism for foreign clients. However, in spite of having 398 medical schools they do not have enough primary care physicians for rural areas. The private sector health organizations provide about 80% of outpatient and 60% of inpatient care, but overall about 70% of healthcare consists of out of pocket spending. India is moving towards universal coverage and as such they passed a draft National Health Policy in 2015. The reality is that universal healthcare would likely cost $23 billion US dollars over the next four years, in order to serve one sixth of the world's population. This is very unlikely given the fact that India only spends about $61 US dollars per capita and healthcare amounts to only 4% of the gross domestic product (GDP). Universal health coverage is not the only priority as India also faces widespread poverty and poor sanitation. India is delivered by a mix of private

and public health services. The country has a wide range of healthcare demands, from very poor rural areas that lack adequately trained healthcare workers, to dense urban areas with booming populations placing a strain on existing healthcare services in the public and private sectors. Although the country has seen rapid economic development in recent years, healthcare provision for the majority of India's inhabitants is still very poor by international standards with high levels of infant mortality, poor sanitation and infectious disease control.[177-179]

National Health Authority (NeHA)
Although EHRs have been adopted by many private hospitals and clinics in India, they are largely focused around billing and patient administration and lack interoperability tools that would enable continuity of care. To address this problem, the government of India produced the "EMR/EHR Standards for India" report and is in the process of setting up a new National eHealth Authority to guide institutional adoption of international standards such as SNOMED-CT.[180]

SMARTHealth
SMARTHealth India is a mHealth initiative from the George Institute for Global Health, based in Sydney Australia, the UK, India, and China. The project uses smartphones and tablet PCs to give community health workers (CHWs) the tools to help diagnose and manage chronic diseases, such as diabetes and heart disease. This means that tasks currently performed by doctors can be "task-shifted" to CHWs while still maintaining high standards of care through the use of clinical decision support and communication tools on the mobile devices.[181]

Sana Mobile
Sana Mobile is a volunteer organization based at the Massachusetts Institute of Technology in the US. They have partnered with a range of organizations including universities, NGOs, health organizations and social enterprises to create a number of projects aimed at using mobile phones to provide healthcare services in LMIC.

The Sana Mobile open source platform allows CHWs to communicate with medical specialists using text, audio, video, and photos for real-time decision support. One of the largest implementations of the Sana Mobile systems has been to help detect oral cancer in rural south India. CHWs can use their phones to take photos of the inside of the patient's mouth and send the images or video to a doctor who can review it and communicate back to the CHW.

The open source Sana platform uses an Android mobile phone application linked to a web interface created using the Django Python Web Framework that is, in turn, linked to the Java OpenMRS open source EMRs.[182]

Singapore

History

Singapore has been forerunner in digitizing national health since their launch of the Patient Master Index (PMI) in 1985 that connected 5 large hospital sites to store and access minimal patient data.[183] By 1995, the next installment was the National Patient Master Index that included limited demographics, allergies, and a small number of specific medical alerts.[184] Singapore announced their intent to create one of the first NEHRs in 2009, and launched the first phase in 2011.[185-186] The current iteration includes analytics for research and online patient access.[187]

In 2015 they launched the multifaceted HealthHub portal, available on PC and mobile.[188] It provides patient access to some health records, appointment booking, health articles, provider/facility directory, and offers deals or rewards at wellness centers, recreation facilities, and food and beverage establishments. Residents access the portal via their Singapore Personal Access or "SingPass"; a government issued account system launched in 2003 for accessing over 600 online eservices.[188-189]

EHR Adoption

Overall use of NEHRs has been ramping up with over 11,400 users and 600,000 record searches in March 2016, a 63% increase in users and a 3-fold increase in searches from the previous year.[190] Adoption of EHRs is higher by public practitioners, pharmacies, and labs. GPs have expressed their hesitance lies in potential interference with daily operations, increased costs, and unfamiliarity with technology.[191]

Interoperability

Singapore's Ministry of Health Holdings has guided interoperable development by establishing the framework for health IT, coordinating continuity with vendors, involving providers early in the process, and engaging in monitoring and management at each phase.[190-191] Connectivity of patient information has also reportedly been less difficult in Singapore as current programs are built on previously established systems.[193]

mHealth

Residents engage in high Internet (70%) and social media (60%) usage driving a blooming mHealth market.[194] Many programs are government and health facility sponsored to track patient health between appointments or for patients to diary and report daily aspects of diet, water intake, and physical activity.[195] There are apps currently under development that will relay health stats for those with conditions such as hypertension or kidney failure, monitor them for abnormalities, and alert providers if intervention is required.[196]

Telemedicine

With a blend of a rapidly aging population and high level of chronic disease, Singapore has been open to exploring options that reduce demand on their finite healthcare resources.[197-198] To this end, they have been gradually adjusting guidelines since the 90's to increase implementation of telehealth in their country.[199-201] While not as rapidly adopted as EHRs, telehealth implementation is ramping up in a variety of specialties: ophthalmology, physical rehabilitation, elder care, and cardiology.[202-205]

Potential Issues

At present, typical patient access is limited to read-only summaries of care but increased patient literacy, use of apps and wearables, and virtual care is driving demand for greater access. There is a growing division between patients and providers with 80% of patients wanting full access while only 17% of doctors reporting the same desire.[206] Some challenges noted in Singapore are also some of the driving forces behind their HIT agenda. The "Silver Tsunami" of increasing elderly, prevalence of chronic disease, and health staff shortage are all aspects that need to be addressed both by and for the advancement of HIT.[207-208]

Bright Future

Singapore has many advantages that have contributed to the successful implementation of HIT including small size, incremental and programmatic sharing of health data, high physician technology literacy, and patient and provider demand for access.[209-210] The country also has the benefit of positive global perception for being a recognized leader in health efficiency, fostering an environment that encourages innovation and expansion of e-health, proper legal structure to protect research and intellectual property, and strong cooperative ties between government and international private industry.[210-212]

Health Informatics in South America

Although a relatively late adopter of HIT internationally, there are now a number of interesting health informatics initiatives occurring across Central and South America.

Brazil

The Brazilian National Health System: A brief review:
In the 1960s and 1970s, Brazilian HISs were mostly distributed across both public and private health organizations. It was after the 1980's that the Brazilian health community established the need for an integrated NHIS, as noted in some of the Brazilian National Health Conferences.[213] The main change was the Brazilian Federal Constitution, which established the rights for public health in 1989 to all Brazilian people, as a fundamental and constitutional right, to be provided by the Government in the form of a Unified Health System (bras. SUS) or national health system.[213-214]

The continental length, the huge regional asymmetries concerning social and technological development, and the coexistence of private and public health entities results in a complex mosaic scenario that has created the main challenges in the Brazilian system.[214-215] Many health information systems have been developed and, in the 1990's, the Informatics Department of SUS, called DATASUS, was created.[216] The continuing social and scientific development as well as the economic growth and stability achieved by Brazil in this period have brought some of the necessary conditions to support the engagement of more ambitious national plans in the context of the decentralized Brazilian health system. In this period, some institutions have recommended changes to national health

information policies, such as the Interagency Web of Information for Health (bras. RIPSA), the National Web of Health Information (bras. RNIS), the Health Informatics Brazilian Society (bras. SBIS) and the National Card for Health project (SUS Card).[217]

Figure 25.6 The Brazilian SUS card

The SUS Card project has been seen as an essential element for the integration and management of Brazilian NHITS throughout the decentralized health entities, similar to the German Card. In 1999 it was implemented as a pilot project with around 13 million people.[218]

The first steps were to register all Brazilians, living in metropolitan and remote regions. Many challenges were found during this prolonged beginning and the methods of implementing the SUS Card program have been continuously reviewed and updated. There are Apple and Android smartphone apps that can access the SUS card. Access to the card enables users to view weight, blood pressure, glucose, allergies, vaccines, doctor visits and medications.[219]

The Brazilian National Plan for Health Information and Informatics
The Brazilian National Policy for Information and Informatics in Health (PNIIS) was endorsed by the United Nations Program for Development.[220]

In 2011, the Health Ministry established the basic standards for interoperability and for health information. The chosen reference model for EHRs was the OpenEHR platform.[221] The architecture of the clinical document is specified by HL7 CDA and SNOMED-CT standards; image results use the DICOM standard and the ISBT 128 standard is used for encoding information related to blood, tissues, and human organs. In 2015 the PNIIS was institutionalized by the

Health Ministry.[222] The principles of the PNIIS includes an individual unified NHIS, with guarantees of health information access at no cost, as well as security and privacy.

Collective Health Data in Brazil

The main HISs of DATASUS are organized into the following categories: 1) National Registers; 2) Outpatient; 3) Epidemiologic; 4) Hospital; 5) Social; 6) Financial; 7) Life Events; and 8) Regulation.[213] These information systems involve a basic healthcare system, outpatient information system, pregnancy tracking system, immunization programs, renal replacement management system, system for registration and monitoring of hypertension and diabetes patients (HIPERDIA), system for management of transplants of organs and tissues, systems for cancer surveillance, epidemiologic data surveillance, and several others.[223-226] DATASUS also provides a national health data warehouse called TABNET, which is open for public access. The interface provides graphical visualizations, using charts and colored maps and statistical analysis.[227]

Preparing people for the use of HIT and its standards in Brazil

Other important guidelines in the PNIIS are those relating to HIT training. The strategy has included efforts from the Ministry of Telecommunications for IT infrastructure and data communication, and from the Ministry of Science and Technology, to stimulate universities and institutes to meet the demand for research and development in HIT.

Finally, the Ministry of Education took part in the national plan, and supports the national distance e-learning project called Brazil Open University (bras. UAB). The UAB courses are mostly supported by Brazilian universities and the Federal Government and Municipalities. E-learning plays an important role for training of SUS staff in Brazil. The Health Ministry created the federal program called UniverSUS to develop and make available free e-learning courses on information and health informatics.[228]

Argentina

Argentina has a rich history of academics, including computer science. However, unlike many countries it has had decades of political instability that dramatically interfered with the evolution of the sciences. Under more democratic circumstances the Latin American School of Higher Education in Informatics was created in 1985 but later closed in 1990 due to insufficient financial support.

Interest in computer science and informatics continued, in spite of setbacks. The program Conectar Igualdad was established to distribute netbook computers to all students ages 10-12 to make them more competitive in technology. To date several million computers have been distributed.[229]

Argentina was the main participant in a new Medical Informatics online course delivered in the 2009-time frame. Oregon Health and Science University partnered with the Hospital Italiano of Buenos Aires to translate and adapt an AMIA 10 x 10 course. Most of the students participating were healthcare professionals from Argentina.[230]

As of 2014 there were 139 private and public institutions of higher education that offer either undergraduate or associate degrees in informatics and 9 doctoral programs.[231]

Uruguay

Uruguay has a national EHR system that was implemented in 2006 under the auspices of the "National Integrated Health System." This system was mandated for all healthcare providers from 2007 onwards and integrated prior private and public partners into one healthcare system.

Uruguay uses HL7v3 CDA documents and IHE profiles to enable interoperability between hospital systems. In addition to their NEHR, they also have a Vital Statistics project, a Maternal and Child Health Program, a Perinatal Information System, the Aduana Program (for child health up to 2 years), and Electronic Death Certificate (CD-e).[232]

In 2008 President Vazquez initiated the Ceibal Project to distribute simple wireless laptops, known as XO computers to every child.[234] This is part of the "One Laptop Per Child" program that has distributed more than 2 million computers in 42 countries.[234]

Bolivia

Text To Change (TTC) is a social enterprise with

Text To Change (TTC) is a social enterprise with offices in Uganda, Amsterdam and Boliva that offers text-messaging programs for a variety of healthcare projects. Examples of their programs include anti-smoking, financial, farming and handwashing educational programs. It now runs campaigns in 17 countries across the world. TTC is using relatively simple, low-cost technology to support public health campaigns that reach many thousands of people.[235]

Health Informatics in North America

Canada

Paving the way for Canadian HIT
In 1994, the National Forum on Health was launched and the recommendation was made to move toward a NHITS. [236] In addition, the Canadian Advisory Council on Health Infostructure began recommending a nationally unified health information system or "information highway" in the late 1990's.[237] To this end, $500 million was budgeted in 2001 to Health Infoway to develop and implement EHRs. During 2003 and 2004 an additional $700 million was granted to increase interoperability, address rural adoption, and develop a Pan-Canadian health surveillance program.[238]

Canada Health Infoway
Canada Health Infoway is an independent, not-for-profit, organization whose membership is derived from deputy ministers of the federal government, territories, and provinces. The board of directors is comprised of a variety of community advisors including those from public health, technology, legal, and finance sectors.[239]

With a substantial investment of public funds, both progress and transparency are paramount. Infoway regularly monitors progress of projects against a "Benefits Evaluation Framework" and publishes both individual and annual reports on overall progress.[240]

The goals of Infoway were to accelerate the implementation of a NEHR and to develop an EHR blueprint for the creation of individual EHRs that suit the needs of differing regions and specialties.[111] Infoway encourages participation and adherence to the Blueprint by reimbursement through a "gated funding" approach, or the reimbursement of up to 75% of

eligible costs for conforming products based on milestones achieved.[241]

Infoway released the first version of the Blueprint in 2003 to deal with data standards and a second version in 2006. Version 2 was released to optimize eHealth and EHR projects by better defining both form and function, expanding into addressing telehealth and disease surveillance, and aligning projects with the Privacy and Security Architecture.[242]

Interoperability and Adoption
Despite having a unified board to oversee development and an architectural blueprint, there has been a lack of interoperability amongst the systems. While Infoway is funding the production of eHealth products, it may not be monitoring them sufficiently. This has led to 11 separate jurisdictions with different standards and different software. The end result has been poor interoperability between the provinces.[243]

By 2014, EHR adoption by primary care physicians had reached an average rate 77 %, 99% for image digitization, 81% for electronic lab test results, and 98% for hospital access to telehealth applications.[239, 244]

Patient Access to Personal Health Records
While 8 in 10 Canadians have expressed a desire to access their health care records and test results online, reportedly only about 6% currently have the ability.[245] To bridge the gap, some regions and specific hospitals are beginning to provide their patients with various online health options, such as RelayHealth, a web-based patient portal.[245]

mHealth
Canadians are gradually beginning to embrace the addition of mHealth into the traditional medical setting. Recent reports indicate that while 74% of Canadian's have accessed health information online only 15% currently use a mHealth app.[246]

An interesting mHealth initiative being promoted is a joint effort between Saint Elizabeth Healthcare and Samsung to provide 5,000 home health workers and 500 administrators with tablets and mHealth apps to manage and monitor patients in variety of home programs from physiotherapy to wound, cancer, and palliative care. Home workers will be able to optimize patient care by tracking the most efficient routes, securely inputting patient data, and scheduling follow up appointments.[247]

Mexico

Mexico's healthcare system has public institutions and private providers, all under the Ministry of Health. The public sector accounts for about 44% of the population while the rest are served by the System for Social Protection in Health, a health insurance program for the poor. Health care is delivered through the 32 state Health Services.[248]

Mexico was the first Latin American country to introduce an EHR on a larger scale at the Mexican Institute of Social Security (IMSS), responsible for administering social security and healthcare benefits. The EHR was introduced in 2003 and consisted of multiple linked databases. A study by Perez-Cuevas et al. demonstrated that this basic EHR was able to evaluate the care of type 2 diabetics using EHR data. However, the study was based on only four clinics and two of the clinics could not provide A1c data.[249]

The state of Colima introduced another EHR system known as SAECCOL in 2005, based on Microsoft SQL server databases and Visual Basic. This was a joint project between the US Agency for International Development (USAID) and the Mexican National Public Health Institute. The modular approach had limitations, such as no inpatient modules. Hernandez-Avila et al. pointed out that many of the EHRs in Mexico were developed prior to an official Mexican EHR standard of 2010 so there is great disparity in terms of functionality and interoperability. His published evaluation of the program in 2012 was not dissimilar from early studies from developed countries. In other words, the same challenges faced by the US (funding, standards, lack of training, privacy, resistance to change, etc.) were experienced in this program.[248]

Resources

eHealth Strategy Toolkit

The World Health Organization (WHO) and the International Telecommunications Union (ITU) have been working to help LMIC achieve higher levels of ICT adoption through the development of eHealth strategies and plans. In 2011, they jointly published the eHealth Strategy Toolkit that gives advice and guidance to developing countries as they begin the process of adopting EHR and other HIS.

Although the toolkit provides good advice on developing a national eHealth vision, it also points out the need to develop a concrete eHealth plan and then to continue to monitor progress through regular assessments and re-evaluations of the plan.[250]

IMIA: International Medical Informatics Association

The International Medical Informatics Association (IMIA) is an organization that brings together health and biomedical informatics organizations from around the world. It describes itself as an "association of associations." Its stated goals are as follows:

- "promote informatics in health care and research in health, bio, and medical informatics.
- advance and nurture international cooperation.
- to stimulate research, development, and routine application.
- move informatics from theory into practice in a full range of health delivery settings, from physician's office to acute and long term care.
- further the dissemination and exchange of knowledge, information, and technology.
- promote education and responsible behaviour.
- represent the medical and health informatics field with the World Health Organization and other international professional and governmental organizations."

IMIA hosts the MedInfo conference series, sponsors the publication of several leading health informatics journals, and hosts a number of Working Groups on a variety of health informatics topics with researchers contributing from around the world.[251]

WHO Atlas of eHealth Country Profiles

In 2015 the WHO published the results of the third global eHealth survey of 125 countries. This is an excellent up-to-date resource for statistics regarding populations, physician and nurse density, hospitals, life expectancy, etc. Sections for each country include: eHealth foundations, legal frameworks for eHealth, telehealth, EHRs, use of eLearning in health sciences, mHealth, social media and big data. [252]

Challenges and Barriers

Lack of Infrastructure
LMIC generally lack the funds to develop high speed Internet access using fiber optic cabling, so they migrate towards mHealth solutions with smartphones. In spite of high penetrance of this technology, it is not scalable for many of the HIT/ICT solutions currently available.

Clinical Adoption
One of the most significant barriers has been clinician buy-in of new systems. Even in environments where systems are provided at no cost to providers, such as the UK's National Project for Information Technology or Canada's Health Infoway, switching from a paper-based workflow to electronic workflow has been difficult.

Interoperability
Interoperability between different EHR systems remains a significant challenge internationally. Countries that have achieved success in exchanging clinical documents, such as Denmark (MedCom) are exceptions, rather than the rule. Issues such as over-ambitious targets for "semantic interoperability" using the HL7 version 3 standard have hampered government-directed standards adoption programs. On the other hand, issues with inconsistent vendor and provider adoption of HL7 version 2 continue to present problems, although several countries have successfully adopted HL7 version 2 for e-Prescribing and e-Discharge Summaries. The new HL7 FHIR standard appears to offer some help with these issues offering a solution that is somewhere between the v2 flexibility and v3 semantic interoperability. Several countries are currently investigating FHIR for national adoption. [253]

Hardware limitations
LMIC struggle to find the resources to fund basic computers and Internet access for the masses. Two exceptions are Ceibal and Conectar Igualidad projects, discussed previously. [233,229]

Software limitations
Adoption of open source solutions has been attractive to LMIC but the reality is that none of these systems are a complete or perfect solution. [254] The OpenEHR framework has been recently adopted by several countries as an attempt to encourage interoperability and accurate data sharing. Although endorsed by the EU (through the CEN 13606 standard), OpenEHR has largely been adopted outside of the EU, with the framework adopted for national registries in Australia, Brazil and Sweden. [221]

Lack of Leadership
Conversion to eHealth/EMR represents a significant paradigm shift, resulting in strained relationships between clinicians and healthcare administrators. Effective leadership is clearly needed to realize a smooth transition. HIS implementation and use is highly knowledge driven and to be successfully implemented, will need support and buy-in from stakeholders both vertically and horizontally. Leaders should be able to motivate and inspire personnel to enhance cooperation and also develop synergies with decisional bodies at local, national, regional, and even global levels. A national eHealth Strategy with consistent funding is mandatory to move forward.

Cost effectiveness
Cost effectiveness of HIT in general has remained a significant issue hampering the adoption of EHRs internationally.

Workforce development
In order to deliver and maintain complex information systems there needs to be a well trained workforce. The AMIA has maintained a Global Partnership Program, funded by the Bill and Melinda Gates Foundation. [255]

Lack of evidence
In 2001 Mitchell said of EMR implementation in resource-limited situations, they are a "descriptive feast but an evaluation famine". [254] The reality is that most articles about HIT/ICT, in all clinical settings, are of low quality, in terms of inadequate number of patients studied, poor

endpoints, poor documentation of unintended consequences, etc.[256] This is covered in more detail in the section on Evidence Based Health Informatics.

Sustainability
Too many HIT/ICT initiatives in LMIC rely on outside funding for "pilot projects", so that when the project concludes there are no funds for sustainment. In the 2014 review by Luna, sustainability was discussed in the light of five necessary factors: effectiveness, efficiency, financial viability, reproducibility, and portability. These are difficult factors in LMIC when faced with civil strife and limited resources.[257]

Conflicting Priorities
In spite of the potential for HIT/ICT initiatives to improve the delivery and documentation of healthcare, it must compete with more basic needs, such as electricity, sanitation, and clean water.

Future Trends

Health Informatics adoption will continue to mature in European countries and other developed economies as they transition to a fully digital provision of health data storage and a range of services for patients and healthcare professionals aimed at making the business of healthcare more efficient and effective. Political and professional issues are likely to continue with areas such as privacy of healthcare information and professional autonomy continuing to be debated as new systems are rolled out.

In the developing world, it is likely that NHIT systems will start being adopted that may favor open source systems over the existing range of North American and European commercial systems. The developing world has been leading the way in the range and scope of mobile device usage and it is likely that these mHealth systems will also mature and become integrated with hospitals and government systems rather than to continue as stand-alone projects largely funded by overseas aid.

International agreements on interoperability standards continue to be a significant issue although new standards, such as HL7 FHIR, may prove to be a breakthrough that enables countries that have adopted a "best of breed" approach to funding EHR systems to share data across the healthcare sector.

Key Points

- High income countries (HIC) and low and middle income countries (LMIC) have adopted substantially different approaches to creating their health information infrastructure

- Mobile technology is playing an important role in health informatics in developing countries

- Cultural and political differences are reflected in the different approaches to national and international health informatics initiatives

- Open source software is an important alternative for LMIC

Conclusion

We are living in an age of rapid adoption of information technology in many industries and healthcare is no exception. By looking internationally at the various approaches and projects taken by different countries it seems that William Gibson was correct in stating that the future is here, but not yet evenly distributed.[258] Some of that future is visible in developing countries in their rapid adoption of mobile technology and some is present in well-funded and coordinate national projects such as the MedCom system in Denmark.

References

1. Schoen C, Osborn R, Squires D et al. A Survey of Primary Care Doctors in Ten Countries Shows Progress in Use of Health Information Technology, Less in Other Areas. Health Affairs. 2012; 31(12):2805-2816

2. McDonald K. Health IT Board working on next five-year plan for NZ. PulseIT. November 16, 2014. http://www.pulseitmagazine.com.au/index.php?option=com_content&view=article&id=2167:health-it-board-working-on-next-five-year-plan-for-nz&catid=49:new-zealand-ehealth&Itemid=274 Accessed October 18, 2015

3. European Commission. Commission publishes four reports of eHealth

    Stakeholder Group. April 11, 2014. https://ec.europa.eu/digital-single-market/en/news/commission-publishes-four-reports-ehealth-stakeholder-group Accessed October 19, 2015

4. Kierkegaard, P. (2013) eHealth in Denmark: A Case Study. Journal of Medical Systems, 37 (6).

5. European Federation for Medical Informatics. www.efmi.org Accessed October 20, 2015

6. I2-Health. Borderless Communication for a Healthy Europe. http://www.i2-health.eu/ Accessed October 19, 2015

7. Austin T, Sun S. Evaluation of ISO EN 13606 As a Result of Its Implementation in XML. Health Informatics Journal. 2013;19(2):264-280

8. Martínez-Costa C, Menárguez-Tortosa M, Fernández-Breis JT. An approach for the semantic interoperability of ISO EN 13606 and OpenEHR archetypes. *J Biomed Inform*. 2010;43(5):736-746.

9. Studzinski J. HIMSS Europe. Three Current EHealth Trends in the German Speaking Region. http://www.himss.eu/node/6916-.VfKu3m1mvSw.linkedin Accessed October 22, 2015

10. Protti D, Johansen I. Widespread adoption of information technology in primary care offices in Denmark: a case study. Commonwealth Fund. March 2010.

http://www.commonweathfund.org/Accessed October 25, 2015

11. 2014 International Profiles of Healthcare Systems. January 2015. Commonwealth Fund. http://www.commonwealthfund.org/Accessed June 7, 2016

12. Civil Registration and Vital Statistics. World Health Organization. 2013. http://www.who.int/healthinfo/civil_registration/crvs_report_2013.pdf Accessed October 20, 2015

13. Sundek. dk https://www.sundhed.dk/ Accessed October 30, 2015

14. Country Brief: Denmark. October 2010. http://www.academia.edu/1400740/Country_Brief_DenmarkAccessed November 3, 2015

15. MedCom. http://medcom.dk/om-medcom Accessed November 1, 2015

16. Kierkegaard, P. Interoperability after deployment: persistent challenges and regional strategies in Denmark. Int. J Quality in Healthcare. 2015. http://intqhc.oxfordjournals.org/content/early/2015/02/25/intqhc.mzv009.full Accessed November 2, 20115

17. Wickland, E. mHealth in Europe: a mixed bag. MobiHealthNews. June 2, 2015. http://www.mobihealthnews.com/Accessed November 10, 2015

18. EU Countries' mHealth App Market. 2015. Research2Guidance. http://www.digitalezorg.nl/digitale/uploads/2015/07/research2guidance-EU-Country-mHealth-App-Market-Ranking-2015.pdf Accessed November 1, 2015

19. Wickland, E. Denmark Kicks Off 5 Telehealth Projects. MobiHealthNews. September 22, 2015. http://www.mobilhealthnews.com/Accessed October 28, 2015

20. Monsenso Aps. http://www.monsenso.com/Accessed November 1, 2015

21. Bennett, J. In Denmark Home Healthcare Rehab working well using Kinect. InternetMedicine. November 24, 2013. http://www.internetmedicine.com/Accessed November 20, 2015

22. Nielson, C, Branebjerg J, Marcussen C, et al. Strategic Intelligence monitoring of personal health systems. Phase 2. Country Study Denmark. European Commission. 2012. http://is.jrc.ec.europa.eu/pages/TFS/documents/SIMPHSCountrystudyDenmarkfinalrev2.pdf Accessed October 22, 2015

23. Executive Summary. Information for Health. An Information Strategy for the Modern NHS 1998-2005. http://webarchive.nationalarchives.gov.uk/+/www.dh.gov.uk/en/Publicationsandstatistics/Publications/PublicationsPolicyAndGuidance/DH_4002944Accessed November 2, 2015

24. Campion-Awwad, Hayton A, Smith L, et al. The National Programme for IT in the NHS. A Case Study. http://www.cl.cam.ac.uk/~rja14/Papers/npfit-mpp-2014-case-history.pdfAccessed October 20, 2015

25. NHS. A Guide to the National Programme for Information Technology. 2005 http://www.providersedge.com/ehdocs/ehr_articles/A_Guide_to_the_National_Programme_for_Information_Technology.pdf Accessed October 30, 2015

26. Dismantling the National Programme for IT. Gov. Uk. September 2011. https://www.gov.uk/government/news/dismantling-the-nhs-national-programme-for-it Accessed November 5, 2015

27. Currie, W. Translating Health IT Policy Into Practice in the UK NHS. Scan J of IS. 2014;26(2):3-26

28. Maughan, A. Six reasons why the NHS National Programme for IT failed. Computer Weekly. http://www.computerweekly.com/opinion/Six-reasons-why-the-NHS-National-Programme-for-IT-failedAccessed October `5, 2015

29. Everybody's Business—strengthening health systems to improve health outcomes. WHO 2007. http://apps.who.int/iris/handle/10665/43918Accessed October 24, 2015

30. Wallace P, Delaney B, Sullivan F. Unlocking the research potential of the GP electronic care record. *Br J Gen Pract.* 2013;63(611):284-285.

31. Gill L, Goldacre M. English national record linkage of hospital episode statistics and death registration records. Unit of Health-Care Epidemiology, Oxford University, Oxford. 2003. http://nchod.uhce.ox.ac.uk/NCHOD Oxford E5 Report 1st Feb_VerAM2.pdf Accessed February 20, 2016

32. Smyth RL. Regulation and governance of clinical research in the UK. *BMJ.* 2011;342:d238

33. Department of Health. Raising the Profile of Long Term Conditions Care A Compendium of Information. Department of Health; 2008. http://webarchive.nationalarchives.gov.uk/20130107105354/http://www.dh.gov.uk/prod_consum_dh/groups/dh_digitalassets/documents/digitalasset/dh_082067.pdfr Accessed February 15, 2016

34. Department of Health. Whole System Demonstrator Programme Headline Findings – December 2011. Department of Health; .http://www.gov.uk/ Accessed February 15, 2016

35. Cartwright M, Hirani SP, Rixon L, et al. Effect of telehealth on quality of life and psychological outcomes over 12 months (Whole Systems Demonstrator telehealth questionnaire study): nested study of patient reported outcomes in a pragmatic, cluster randomised controlled trial. *BMJ.* 2013;346:f653.

36. Department of Health: The National Programme for IT in the NHS: 2006-2007. http://www.publications.parliament.uk/pa/cm200607/cmselect/cmpubacc/390/390.pdf Accessed November 11, 2015

37. Greenhalgh T, Stramer B, Bratant T et al. The Devil's in the Details. 7 May 2010 https://www.ucl.ac.uk/news/scriefullreport.pdfAccessed October 28, 2015

38. De Lusignan S, Seroussi B. A Comparison of English and French approaches to providing access to summary care records: scope, consent, cost. Stud Health Tech Inform 2013;186:61-65

39. NHS. Personalised Health and Care 2020. November 2014. https://www.gov.uk/government/uploads/system/uploads/attachment_data/file/384650/NIB_Report.pdfAccessed November 1, 2015

40. Thompson, S. International Profiles of Healthcare Systems. 2013. Commonwealth Fund. http://www.commonwealthfund.org/Accessed October 15, 2015

41. NHS. Five Year Forward View. October 2014. http://www.england.nhs.uk/wp-content/uploads/2014/10/5yfv-web.pdf Accessed October 25, 2015

42. Gov. UK. Plans to improve digital services for the health and care sector. National Information Board. 19 June 2015. http://www.gov.uk/Accessed October 20, 2015

43. NHS. Interoperability Handbook. September 2015. http://www.england.nhs.uk/ Accessed November 15, 2015

44. Monegain, B. Epic EHR adds to UK hospital's financial mess. HealthcareITNews. September 28 2015. http://www.healthcareitnews.com/Accessed November 12, 2015

45. iPlato. http://www.iplato.net/Accessed December 1, 2015

46. Tunstall Healthcare. www. Tunstall.co.uk Accessed December 1, 2015

47. Personaltechmd. http://www.personaltechmd.com/Accessed December 1, 2015

48. Raghavulu, V. Prasad, A. Role of computer science in healthcare. Int J Sci Engin Tech Res 2014;3(46):9386-9387

49. Currie W, Seddon J. A cross-national analysis of eHealth in the EU: some policy and research directions. Inf Man 2014;51(6):783-797

50. Stroetmann KA, Artmann J, Giest S. Country Brief: Germany. October 2010. eHealth Strategies http://www.ehealth.strategies.eu/Accessed November 15, 2015

51. Tuffs A. Germany plans to introduce electronic health card. BMJ. 2004;329:7458

52. Hoeksma J. Germany suspends ehealth card project. Digital Health. 19 January 2010. http://www.digitalhealth.net/ Accessed October 28,2015

53. Germany and the challenge of rolling out eHealth on a large scale. 15 May 2013. http://www.esante.gour.fr/Accessed November 15, 2015

54. The Health Systems and Policy Monitor. Germany. 2014. http://www.hspm.org/countries/GermanyAccessed October 28, 2015

55. Viactiv. https://www.viactiv.de/english/electronic-health-card-egk/ Accessed October 28, 2015

56. The Federal Government. Digital Agenda 2014-2017. http://www.digitale-agenda.de/ Accessed November 20, 2015

57. European Commission. Hillenius G. Germany's digital agenda reshuffles country's eHealth policy. January 9, 2015. https://joinup.ec.europa.eu/Accessed November 20, 2015

58. Federal Ministry. Act on secure digital communication and applications in the healthcare system. September 29, 2015. http://www.bmg.bund.de/ Accessed October 20, 2015

59. Bertlemann H, Dinger F, Schreiber L. Germany: New healthcare reforms 2015. Health Law Pulse. August 19, 2015. http://www.thehealthlawpulse.com/Accessed November 1, 2015

60. Germany Trade and Invest. German mHealth market outlook. http://www.gtai.de/Accessed October 25, 2015

61. Sesam Vitale. Smart Card Alliance. http://d3nrwezfchbhhm.cloudfront.net/pdf/Sesam_Vitale.pdfAccessed December 5, 2015

62. Overview of the national laws on EHR member states. National report for France. Milieu Law and Policy Consulting. January 2014. http://ec.europa.eu/health/ehealth/docs/laws_france_en.pdfAccessed December 1, 2015

63. De Lusignan S, Ross P, Shifrinm P, Seroussi B. Comparison of approaches to providing patients access to summary care records across old and new Europe: an exploration of facilitators and barriers to implementation. Stud Health Tech Inform 2013;192(1):397-401

64. Dossier Medical Personnel. http://www.dmp.gouv.fr/web/dmp/ Accessed November 15, 2015

65. Sesam-Vitale Evaluation. A contribution to French sustainability policy. Gemalto. http://www.gemalto.com/brochures-site/download-site/Documents/gov_sesam_vitale.pdf Accessed November 5, 2015

66. What is a CPS card? 2014 http://esante.gouv.fr/en/services/espace-cps/what-a-cps-card Accessed November 4, 2015

67. Health Informatics Systems Interoperability Framework (HIS-IF). 6 May 2015. http://esante.gouv.fr/en/node/2053 Accessed November 20, 2015

68. Gray B, Bowden T, Johansen I et al. Electronic Health Records: An International Perspective on "Meaningful Use". The Commonwealth Fund November 2011. http://www.commonwealthfund.org/Accessed November 20, 2015

69. Park Y, Atalag K. Current National Approach to Healthcare ICT Standardization: Focus on the Progress in New Zealand. Health Inform Res 2015;21(3):144-151

70. McDonald K. HiNZ 2015: National child health IT platform scores early wins. Pulse IT. 22 October 2015. http://www.pulseitmagazine.com.au/ Accessed November 30, 2015

71. OrionHealth. https://orionhealth.com/Accessed November 24, 2015

72. Ministry of Health. New Zealand Health Strategy 2016. http://www.health.govt.nz/ Accessed June 13, 2016

73. Smartphones changing the face of mobile health. National Health IT Board. 9 April 2015. http://www.healthitboard.health.govt.nz/Accessed December 15, 2015

74. Beating the Blues. http://www.beatingtheblues.co.nz/Accessed December 15, 2015

75. Bartlett C, Boehncke K. E-health: enabler for Australia's health reform. November 2008. Booz & Co.

76. Healthcare Identifiers Act of 2010. http://www.legislation.gov.au/ Accessed December 2015

77. Australian Government Department of Health. Personally Controlled EHR for all Australians. 14 September 2012. http://www.health.gov.au/ Accessed December 11, 2015

78. McDonald K. Problems plaguing PCEHR provider portal. Pulse IT. 26 February 2014. www.pulseitmagazine.com.au Accessed December 20, 2015

79. My Health Record. https://myhealthrecord.gov.au/Accessed June 13, 2016

80. Is Australia ready for mobile health? Healthcare Innovation. August 18, 2015. http://www.enterpriseinnovation.net/Accessed December 20, 2015

81. Telstra. http://www.telstra.com.au/ Accessed January 2, 2016

82. Personal KineticGraph. Global Kinetics Corporation. http://www.globalkineticscorporation.com/ Accessed January 2, 2016

83. Van de Maele N, Evans D, Tan-Torres T. Development assistance for health in Africa: are we telling the right story? Bulletin for the WHO. 2013;91:483-490

84. Sharma P, Agarwal P. Mobile phone text messaging for promoting adherence to antiretroviral therapy in patients with HIV infection. The WHO Reproductive Health Library; Geneva: World Health Organization 2012(3)

85. Department of State, 2012. *PEPFAR Blueprint :Creating an AIDS-free Generation,*

86. Frasier, H., May, M.A. & Wanchoo, R., 2008. *e-Health Rwanda Case Study,* Available at: http://ehealth-connection.org/files/resources/Rwanda + Appendices.pdf.

87. Afarikumah E. Electronic health in ghana: Current status and future prospects. Online J. Public Health Inform 2014, May;5(3):230

88. Were MC, Siika A, Ayuo PO, Atwoli L, Esamai F. Building Comprehensive and Sustainable Health Informatics Institutions in Developing Countries: Moi University Experience. Studies in health technology and informatics. 2015;216:520.

89. Manya A, Braa J, Øverland LH, et al. National roll out of district health information software (DHIS 2) in Kenya, 2011--central server and cloud based infrastructure. http://www.ist-africa.org/Conference2012

90. Braa J, Humberto M. Building collaborative networks in Africa on health information systems and open source software development--experiences from the HISP/BEANISH http://www.ist-africa.org/Conference2007

91. Wolfe BA. The openmrs system: Collaborating toward an open source EMR for developing countries. AMIA Annu. Symp. Proc 2006:1146

92. OpenMRS. http://www.openmrs.org/Accessed March 23, 2016

93. Levy S. Zuckerberg explains internet. org, Facebook's plan to get the world online. August 26, 2013. http://www.wired.com/ Accessed May 14, 2016

94. SpaceX founder files with government to provide Internet service from space. The Washington Post. 2015 http://www.washingtonpost.com/business/economy/spacex-founder-files-with-government-to-provide-internet-service-from-space/2015/06/09/db8d8d02-0eb7-11e5-a0dc-2b6f404ff5cf_story.html. Accessed May 14, 2106

95. Handwerk B. Google's Loon Project Puts Balloon Technology in Spotlight. *Natl Geogr Mag*. 2013.

96. The World in 2014. http://www.itu.int/en/ITU-D/Statistics/Documents/facts/ICTFactsFigures2014-e.pdf Accessed May 14, 2016

97. Medic Mobile. http://medicmobile.org/Accessed June 10, 2016.

98. FrontlineSMS | FrontlineCloud. http://www.frontlinesms.com/ Accessed June 10, 2016.

99. GAVI. http://www.gavi.org/Accessed June 10, 2016

100. Gerber T, Olazabal V, Brown K, Pablos-Mendez A. An agenda for action on global e-health. *Health Aff* . 2010;29(2):233-236.

101. Saxenian H, Cornejo S, Thorien K, et al. An analysis of how the GAVI alliance and low- and middle-income countries can share costs of new vaccines. *Health Aff*. 2011;30(6):1122-1133.

102. Vodafone. http://www.vodafone.com/Accessed June 4, 2016

103. Roll Back Malaria http://www.rollbackmalaria.org/ Accessed May 10, 2016

104. Githinji S, Kigen S, Memusi D, Nyandigisi A, Mbithi AM, Wamari A, Muturi AN, Jagoe G, Barrington J, Snow RW, Zurovac D. Reducing stock-outs of life saving malaria commodities using mobile phone text-messaging: SMS for life study in Kenya. PLoS One. 2013 Jan 17;8(1):e54066

105. Siddle K, Vieren L, Fiander A. Characterising women with obstetric fistula and urogenital tract injuries in Tanzania. Int Urogynecol J Pelvic Floor Dysfunct. 2014;25(2):249–55.

106. M-pesa. http://www.safaricom.co.ke/personal/m-pesa Accessed May 10, 2016

107. Coleman J. Monitoring MAMA: Gauging the Impact of MAMA South Africa. *J Mob Technol Med*. 2013;2(4s):9.

108. Grice HP, Cole P, Morgan JL. Syntax and semantics. Logic and conversation. 1975;3:41-58

109. Rothman AJ, Salovey P, Antone C, Keough K, Martin CD. The Influence of Message Framing on Intentions to Perform Health Behaviors. *J Exp Soc Psychol*. 1993;29(5):408-433

110. Kerrison RS, Shukla H, Cunningham D, et al. Text-message reminders increase uptake of routine breast screening appointments: a randomised controlled trial in a hard-to-reach population. *Br J Cancer*. 2015;112(6):1005-1010.

111. de Jongh T, Gurol-Urganci I, Vodopivec-Jamsek V, et al. Mobile phone messaging for

facilitating self-management of long-term illnesses. *Cochrane Database Syst Rev.* 2012;(12):CD007459.pub2.

112. Gurol-Urganci I, de Jongh T, Vodopivec-Jamsek V, Atun R, Car J. Mobile phone messaging reminders for attendance at healthcare appointments. Cochrane database Syst Rev 2013. http://www.ncbi.nlm.nih.gov/pubmed/243 10741 Accessed June 1, 2016

113. Car J, Gurol-Urganci I, De Jongh T, Vodopivec-Jamsek V, Atun R. Mobile phone messaging reminders for attendance at scheduled healthcare appointments. *Cochrane Database of Systematic Reviews.* 2008;(4). doi:10.1002/14651858.CD007458.

114. Shet A, De Costa A, Kumarasamy N, et al. Effect of mobile telephone reminders on treatment outcome in HIV: evidence from a randomised controlled trial in India. *BMJ.* 2014;349:g5978.

115. Pop-Eleches C, Thirumurthy H, Habyarimana JP, et al. Mobile phone technologies improve adherence to antiretroviral treatment in a resource-limited setting: a randomized controlled trial of text message reminders. *AIDS.* 2011;25(6):825-834.

116. Ogara E, Magana O. eHealth Strategy 2011 - 2017. 2011. http://www.isfteh.org/files/media/kenya_n ational_ehealth_strategy_2011-2017.pdf Accessed June 8, 2016

117. Vision 2030. http://www.vision2030.go.ke/Accessed June 10, 2016

118. MOH. Kenya EMR Review Towards Standardization Report. September 7, 2011. https://www.ghdonline.org/ Accessed June 11, 2016

119. West D, Branstetter DG, Nelson SD, Manivel JC, Blay J-Y, Chawla S, et al. How Mobile Devices are Transforming Healthcare. BrookingsEdu 2012;18(16):1–38

120. Sema Doc. http://hellodoctor.co.ke/Accessed June 10, 2016

121. Piette J, Lun K, Moura L, et al. Impacts of e-health on the outcomes of care in low- and middle-income countries: where do we go

from here? Bull World Health Organ. 2012;90(5):365-372.

122. Ministry of ICT. The Kenya National ICT MasterPlan : Towards a Digital Kenya. 2014. https://www.kenet.or.ke/sites/default/files/ Final ICT Masterplan Apr 2014.pdf Accessed June 10, 2016

123. Nisingizwe MP, Iyer HS, Gashayija M, et al. Toward utilization of data for program management and evaluation: quality assessment of five years of health management information system data in Rwanda. Global health action. 2014;7

124. Department of Health South Africa, 2012. National eHealth Strategy, South Africa 2012/13-2016/17. *Department of Health, Republic of South Africa*, pp.1–36

125. Springer, 2012. *Foundations of Health Informatics Engineering and Systems* Z. Liu & A. Wassyng, eds., Berlin, Heidelberg: Springer Berlin Heidelberg

126. NISR, 2015. NATIONAL INSTITUTE OF STATISTICS OF RWANDA March 2015 Gross Domestic Product – 2014 Gross Domestic Product and its structure In 2014 , (March), pp.1–8

127. Farmer, P.E. et al., 2013. Reduced premature mortality in Rwanda: lessons from success. *Bmj*, 346(jan18 1), pp.f65–f65.

128. Crichton, R. et al., 2013. An architecture and reference implementation of an open health information mediator: Enabling interoperability in the Rwandan health information exchange. *Lecture Notes in Computer Science (including subseries Lecture Notes in Artificial Intelligence and Lecture Notes in Bioinformatics)*, 7789 LNCS, pp.87–104.

129. UNECA. MDG Report 2014. *Assessing Progress in Africa toward the Millennium Development Goals.* www.undp.org Accessed June 4, 2016

130. Perry, H. *Case Studies of Large-Scale Community Health Worker Programs.* Appendix. www.mchip.net Accessed June 4, 2016

131. Frasier H, May M, Wanchoo R. e-Health Rwanda case study. Am Med http://ehealth-connection.org/files/resources/Rwanda + Appendices.pdf Accessed May 15, 2016

132. Ventures Africa & Iruobe, E., 2015. Rwanda tasked with pioneering eHealth for East Africa. http://venturesafrica.com/rwanda-tasked-with-pioneering-ehealth-for-east-africa/ Accessed June 8, 2016

133. Perry HB, Zulliger R, Rogers MM. Community health workers in low-, middle-, and high-income countries: an overview of their history, recent evolution, and current effectiveness. Annual review of public health. 2014 Mar 18;35:399-421.

134. Aranda-Jan CB, Mohutsiwa-Dibe N, Loukanova S. Systematic review on what works, what does not work and why of implementation of mobile health (mHealth) projects in Africa. BMC public health. 2014 Feb 21;14(1):188

135. Källander K, Tibenderana JK, Akpogheneta OJ, et al. Mobile health (mHealth) approaches and lessons for increased performance and retention of community health workers in low-and middle-income countries: a review. Journal of medical Internet research. 2013;15(1):e17.

136. Kizito K, Adeline K, Baptiste K, Anita A. TRACnet: A National Phone-based and Web-based Tool for the Timely Integrated Disease Surveillance and Response in Rwanda.2013 http://www.ncbi.nlm.nih.gov/pmc/articles/PMC3692857/ Accessed May 21, 2016

137. United Nations, 2007. *TRACnet , Rwanda : Fighting Pandemics through Information Technology.* www.un.org Accessed June 5, 2016

138. Svoronos T, Jillson IA, Nsabimana MM. TRACnet's absorption into the Rwandan HIV/AIDS response. International Journal of Healthcare Technology and Management. 2008 Jan 1;9(5-6):430-45.

139. Leuchowius K. Report on the health care sector and business opportunities in Rwanda. 2014 http://www.swecare.se/Portals/swecare/Documents/Report-on-the-Health-Care-Sector-and-Business-Opportunities-in-Rwanda-Sep2014-vers2.pdf Accessed May 22, 2016

140. Sun R, Cao H, Zhu Z et al. Current aging research in China. Protein Cell 2015;6(8):314-321

141. CHIMA. The white paper on China's hospital information systems. May 2008. http://cdn.medicexchange.com/images/whitepaper/chinas_hospital_information_systems.pdf?1294036467   Accessed January 5, 2016

142. Walton G. China's Golden Shield: corporations and the development of surveillance technology in China. Rights and Democracy. 2001

143. Grace Yu. "China HIT Case Study." in *Health Information and Technology and Policy Lab HIT Briefing Book*, ed. Claire Topal and Kaleb Brownlow (Seattle, WA: National Bureau of Asian Research, 2007)

144. Lei J, Sockolow P, Guan P et al. A comparison of EHRs at two major Peking university hospitals in China to US meaningful use objectives. BMC MI and Decision Making. 28 August 2013. http://www.bmcmedicinformdecismak.biomedcentral.com/

145. Parikh H. Overview of EHR systems in BRIC nations. Clinical leader. April 15, 2015. http://www.clinicalleader.com/ Accessed September 22, 2016.

146. He P, Yuan Z, Liu G et al. An evaluation of a tailored intervention on village doctors use of electronic health records. BMC Health Serv Res. 2014. 14:217

147. International Health Care System Profiles. http://international.commonwealthfund.org/features/ehrs/ Accessed January 15, 2016

148. Xiaohui Y, Han H, Jiadong D et al. mHealth in China and the United States. Center for Tech Innovations at Brookings. http://www.brookings.edu/~/media/research/files/reports/2014/03/12-mHealth-china-united-states-health-care/mHealth_finalx.pdfAccessed January 20, 2016

149. Yoo E. Diabetes management platform Weitang raises series B from Yidu cloud. Tech Node. January 4, 2016. http://www.technode.com/ Accessed January 20, 2016

150. Custer C. Chinese smartphone glucometer DNurse raises millions. Tech In Asia. February 10, 2015. http://www.techinasia.com/ Accessed January 20, 2016

151. Chen L, Wang N, Du X et al. Effectiveness of a smartphone app on improving immunization of children in rural Sichuan Province, China: study protocol for a paired cluster randomized controlled trial. BMC Public Health. 2014. DOI:10.1186/1471-2458-14-262

152. China Digital Health Boot Camp. http://scpku.fsi.standford.edu/Accessed January 25, 2016

153. Zhihua L. Healthcare at your fingertips. China Daily. February 22, 2016. http://www.chinadaily.com.cn/Accessed February 25, 2016

154. Cusano D. Can digital health solve China's healthcare quality and distribution problems? TeleCare. July 23,2015. http://www.telecareaware.com/ Accessed January 20, 2016

155. Sun J, Guo Y, Wang X et al. mHealth for aging China: opportunities and challenges. Aging and Disease. 2016;7(1):53-67

156. Jourdan A. Digital doctors: China sees tech cure for healthcare woes. Reuters. October 14, 2014. http://www.reuters.com/ Accessed January 20, 2016

157. E-Commerce Security: Advice From Experts. M. Khosrow-Pour Editor. Cybertech Publishing. 2004

158. Traditional chine medicine hospitals growing. The State Council. The Peoples Republic of China. January 14, 2016. http://English.gov.cn Accessed February 3, 2016

159. Hsu W, Chan E, Zhang Z et al. A survey to investigate attitudes and perceptions of Chinese medicine professionals in HIT in Hong Kong. Eur J Int Med 2015;7(1):36-46

160. Richie J. China's EMR Market. EMR & HIPAA. October 13, 2013 http://www.emrandhipaa.com/Accessed February 5, 2016

161. Lei J, Xu L, Meng Q et al. The current status of usability studies of IT in China: a systematic review.2014. http://www.hindawi.com/journals/bmri/2014/568303/ Accessed February 5, 2016

162. Zhang Y, Tang Z. Health technology assessment in China. Health Affairs. 2013;32(2):438

163. Abraham C, Nishihara E, Akiyama. Transforming healthcare with information technology in Japan: A review of policy, people, and progress. Int J Med Inform. 2011; 80(3): 157-170.

164. Okada M, Yamamoto K, Kawamura T. Health and medical Informatics education in Japan. IMIA Yearb Med Inform. 2004; 193-198.

165. Sonoda T. Evolution of electronic medical record solutions. Fujitsu Sci Tech J. 2011; 47(1): 19-27.

166. Juhr M, Haux R, Suzuki T, Takabayaski K. Overview of recent trans-institutional health network projects in Japan and Germany. J Med Syst. 2015; 39(5): 50.

167. Ezaki H. President's message. National Hospital Organization Nagasaki Medical Center. http://www.nagasaki-mc.jp/content/en/

168. Organization for Economic Co-operation and Development. OECD health policy studies: Strengthening health informatics infrastructure for health care quality governance. Paris, France: OECD Publishing; 2013

169. Yoshida Y, Imai T, Ohe K. The trends in ENR and CPOE adoption in Japan under the national strategy. Int J Med Inform. 2013; 82(10): 1004-1011

170. Yokobori Y, Sakai T, Takeda T, Oi T. Post Earthquake Health Information Management in Japan—the Challenges. Perspectives in Health Information Management. 2015; 1: 1-13.

171. Matsuda R. The Japanese health care system. In: Mossialos E and Wenzl M, ed. 2015 International profiles of health care systems. New York, NY: The Commonwealth Fund; 2016: 107-114.

172. Essany M. mHealth: Ample room for growth in Japan. mHealthwatch. http://mhealthwatch.com/mhealth-ample-room-for-growth-in-japan-19265/. Published January 3, 2013.

173. Montgomery S. 8 Examples of how mHealth is increasing access to healthcare around the world. Nuviun Digital Health. http://nuviun.com/content/mHealth. Published February 20, 2014.

174. Apple Press. Japan Post group, IBM and Apple deliver iPads and custom apps to connect elderly in Japan to services, family and community. *Apple.* https://www.apple.com/pr/library/2015/04/30Japan-Post-Group-IBM-and-Apple-Deliver-iPads-and-Custom-Apps-to-Connect-Elderly-in-Japan-to-Services-Family-and-Community.html. Published April 30, 2015.

175. Japan Times Staff. Medical industry signals interest in telemedicine. *Japan Times.* http://www.japantimes.co.jp/news/2016/03/16/national/social-issues/medical-industry-signals-interest-telemedicine/#.V2wZUPkrLAW. Published March 16, 2016.

176. Abd Ghani MK, Bali RK, Marshall IM, Wickramasinghe NS. Electronic health records approaches and challenges: A comparison between Malaysia and four East Asian countries. *Int J Electron Healthc.* 2008; 4(1): 78-104.

177. Reddy KS. India's Aspirations for Universal Health Coverage. NEJM 2015;373:1-5

178. Healthcare Industry in India. India Brand Equity Foundation. www.ibef.org Accessed September 23, 2016

179. Deloitte. *2016 Global health care outlook: Battling costs while improving care.* London: Deloitte Touche Tohmatsu; 2016: 1-28.

180. Ministry of Health and Family Welfare. EHR Standards for India. August 2013. http://www.mohfw.nic.in/showfile.php?lid=1672 Accessed January 20, 2016

181. Smart Health India. The George Institute. http://www.georgeinstitute.org/philanthropic-opportunities/smart-health-india Accessed June 25, 2016

182. Sana. http://sana.mit.edu Accessed May 1, 2016

183. Ooi LC, Tan P. Out IT journey: One patient-One record. In: Earn LC, Satku K, eds. *Singapore's Health Care System: What 50 Years Have Achieved.* Hackensack, NJ: World Scientific Publishing Co; 2016: 337-350.

184. Tan, LT. National Patient Master Index in Singapore. *Int J Biomed Comput.* 1995; 40(2): 89-93.

185. Accenture Staff. Accenture implements nationwide electronic health record system in Singapore. Accenture Newsroom. https://newsroom.accenture.com/industries/health-public-service/accenture-implements-nationwide-electronic-health-record-system-in-singapore.htm. Published June 20, 2011. Accessed July 10, 2016.

186. U.S. International Trade Administration. Singapore Type of market: Moderate/Growing. In: *2015 ITA Health IT Top Markets Report.* 2015: 31-34.

187. Liu C, Haseltine W. The Singapore health care system. In: Mossialos E and Wenzl M, ed. 2015 *International profiles of health care systems.* New York, NY: The Commonwealth Fund; 2016: 143-151.

188. Singapore Ministry of Health. Fact sheet on HealthHub. *Ministry of Health.* https://www.moh.gov.sg/content/dam/moh_web/PressRoom/20151018_HealthHub%20Factsheet.pdf. Published October 18, 2015. Accessed July 10, 2016.

189. Singapore Government Staff. eCitizen: eServices. *eCitizen.gov.* https://www.ecitizen.gov.sg/eServices/Pages/Default.aspx#tabs-3. Published 2016. Accessed July 10, 2016.

190. Singapore Ministry of Finance. MOF Committee of Supply speech 2016 by Senior Minister of State for Finance MS Sim Ann. *Ministry of Finance.* http://www.mof.gov.sg/news-reader/articleid/1628/parentId/59/year/2016. Published April 16, 2011. Accessed July 10, 2016.

191. Song, GGY. GPs utilization of NEHR. Ministry of Health. https://www.moh.gov.sg/content/moh_web/home/pressRoom/Parliamentary_QA/2015/gps-utilisation-of-nehr.html. Published February 12, 2015. Accessed July 10, 2016.

192. Shehadi R, Tohme W, Bitar J, Kutty S. *Anatomy of an eHealth ecosystem.* Beirut: Booz&co; 2011.

193. Accenture. Connected health: The drive to integrated healthcare delivery. http://www.himss.eu/sites/default/files/Accenture-Connected-Health-Global-Report-Final-Web.pdf. Published 2012. Access July 10, 2016.

194. Deloitte. *Healthcare and life sicences predictions 2020: A bold future*. London: Deloitte Touche Tohmatsu; 2015: 1-32.

195. Goh G, Tan NC, Malhotra R, et al. Short-term trajectories of use of a caloric-monitoring mobile phone app among patients with Type 2 diabetes mellitus in a primary care setting. *J Med Internet Res*. 2015; 17(2): e33.

196. Wee L. Dr. apps on call 24/7. *Singapore General Hospital*. http://www.sgh.com.sg/about-us/newsroom/News-Articles-Reports/Pages/DrAppsoncall247.aspx. Published March 22, 2012. Accessed July 10, 2016.

197. Chin CW, Phua KH. Long-term care policy: Singapore's experience. *J Aging Soc Policy*. 2016; 28(2): 113-129.

198. HIMSS Asia Pacific. Transforming health through IT. HIMSS Asia Pac. http://www.himssasiapac.org/sites/default/files/HIMSSAP_ExclusiveArticles_CaseStudy_ContinuityofCareforElderly.pdf. Published 2015. Accessed July 10, 2016.

199. Singapore Ministry of Health. National Telemedicine Guidelines. *Ministry of Health*. https://www.moh.gov.sg/content/dam/moh_web/Publications/Guidelines/MOH%20Cir%2006_2015_30Jan15_Telemedicine%20Guidelines%20rev.pdf. Published January 2015. Accessed July 10, 2016.

200. Singapore Medical Council. Ethical code and ethical guidelines. *Health Professionals*. http://www.healthprofessionals.gov.sg/content/dam/hprof/smc/docs/publication/SMC%20Ethical%20Code%20and%20Ethical%20Guidelines.pdf. Published July 12, 2011. Access July 10, 2016.

201. Kwang TW. Smart hospitals, telehealth and EHR in Singapore. *Enterprise Innovation*. http://www.enterpriseinnovation.net/article/smart-hospitals-telehealth-and-ehr-singapore-1526708088. Published September 28, 2015 Accessed July 10, 2016.

202. Hassan NJ. More needs to be done to ensure telemedicine is used in right setting: Experts. *Channel News Asia*. http://www.channelnewsasia.com/news/singapore/more-needs-to-be-done-to/2441808.html. Published January 20, 2016. Accessed July 10, 2016.

203. Senthilingam M, Stevens A. The doctor will not see you now: How Singapore is pioneering telemedicine. CNN. Published November 3, 2015. Accessed July 10, 2016.

204. Wee L. See your doctor over a webcam. Singapore General Hospital. http://www.sgh.com.sg/about-us/newsroom/News-Articles-Reports/Pages/Seeyourdoctoroverawebcam.aspx. Published February 23, 2012. Accessed July 10, 2016.

205. Riistama J., Pauws S, Tesanovic A., et al. First of a kind telehealth implementation study in Singapore: Methodology and challenge to tailor to Asian healthcare system. *Circ Cardiovasc Qual Outcomes*. 2015; 8: A370.

206. Accenture. Consumers and doctors in Singapore increasingly divided on who should have access to a patient's electronic health record, Accenture survey finds. *Accenture*. Published May 18, 2016. Accessed July 10, 2016.

207. Association of Chartered Certified Accounts. Global perspectives on health challenges. ACCA Global. http://www.accaglobal.com/content/dam/acca/global/PDF-technical/health-sector/tech-tp-gpohc.pdf. Published 2012. Accessed July 10, 2016.

208. Rowlands D. Guest editorial: National eHealth record systems – the Singapore experience. *Pulse IT Magazine*. http://www.pulseitmagazine.com.au/asia-pacific-health-it/1139-guest-editorial-national-ehealth-record-systems-the-singapore-experience. Published September 10, 2012. Accessed July 10, 2016.

209. National eHealth Transition Authority Ltd. *Evolution of eHealth in Australia: Achievements, lessons, and opportunities*. Sydney: NEHTA; 2016.

210. Greene W. Health tech innovation blooms in Singapore. Techonomy. http://techonomy.com/2016/04/health-tech-innovation-blooms-in-singapore/. Published April 18, 2016. Accessed July 10, 2016.

211. Lawrence S. Philips, Singapore to jointly invest in population health management companies targeting Asia. *Fierce Biotech*. http://www.fiercebiotech.com/medical-

devices/philips-singapore-to-jointly-invest-population-health-management-companies. Published January 13, 2016. Accessed July 10, 2016.

212. Bloomberg Briefs Staff. 2015 most efficient health care (Table). *Bloomberg Briefs.* http://www.bloombergbriefs.com/content/uploads/sites/2/2015/11/health-care.pdf. Published 2015. Accessed July 10, 2016.

213. Morais RM, Costa AL, Gomes EJ. Information Systems Sus: a Historical Perspective and Policies of Computing and Information. Nucleus. 2014;11(1):239–56

214. Paim J, Travassos C, Almeida C, et al. The Brazilian health system: History, advances, and challenges. Lancet. 2011;377(9779):1778–97

215. Victora CG, Barreto ML, do Carmo Leal M, et al. Health conditions and health-policy innovations in Brazil: the way forward. Lancet. 2011;377(9782):2042–53.

216. Presidency of the Republic. http://www.planalto.gov.br/ccivil_03/decreto/1990-1994/D0100.htm

217. Duarte LS, Pessoto UC, Guimarães RB, et al. Regionalization of health in Brazil: an analytical perspective. Saúde e Soc. SciELO Brasil; 2015;24(2):472–85.

218. Cunha RE Da. Cartão Nacional de Saúde: os desafios da concepção e implantação de um sistema nacional de captura de informações de atendimento em saúde. Cien Saude Colet. 2002;7(4):869–78.

219. Health Ministry launches digital version of SUS card. Portal Saude. www.portalsaude.saude.gov.br Accessed June 17, 2016

220. Brazilian Health Ministry. {PORTARIA} {N} 2073 de Agosto 2011 do Ministério da Saúde. 2011. p. 4.

221. OpenEHR. www.openehr.org Accessed June 1, 2016

222. Brazilian Health Ministry. Brazilian National Plan for Health Information and Informatics - {PNIIS}. DOU -Brazilian Fed Press - Minist Heal. 2015;Portaria 9(Seção 1):72.

223. Prestes IV, de Moura L, Duncan BB, Schmidt MI. A national cohort of patients receiving publicly financed renal replacement therapy within the Brazilian Unified Health System. *Lancet.* 2013;381:S119.

224. Bittencourt SA, Camacho LAB, do Carmo Leal M. O Sistema de Informação Hospitalar e sua aplicação na saúde coletiva Hospital Information Systems and their application in public health. *Cad Saúde Pública.* 2006;22(1):19-30.

225. Junior CN, Padoveze MC, Lacerda RA. Governmental surveillance system of healthcare-associated infection in Brazil. *Rev Esc Enferm USP.* 2014;48(4):657-662.

226. Pires MRGM, Gottems LBD, Vasconcelos Filho JE, Silva KL, Gamarski R. The Care Management Information System for the Home Care Network (SI GESCAD): support for care coordination and continuity of care in the Brazilian Unified Health System (SUS). *Cien Saude Colet.* 2015;20(6):1805-1814.

227. TABNET http://datasus.saude.gov.br/informacoes-de-saude/tabnet Accessed June 2, 2106

228. UNIVERSUS. http://universus.datasus.gov.br/ Accessed June 1, 2016

229. Watkins SC. Conectar Igualdad. Argentina's bold move to build an equitable digital future. 9/16/2011. DML Central. http://dmlcentral.net/conectar-igualdad-argentina-s-bold-move-to-build-an-equitable-digital-future/?variation=b&utm_expid=42725708 5.oNX7Gn2HRIOm8WhS1GUvRg.1&utm_r eferrer=https%3A%2F%2Fwww.google.com %2F Accessed May 20, 2016

230. Otero P, Hersh W, Luna D. A Medical Informatics Distance Learning Course for Latin America. Methods Inf Med. 2010;49:310–5

231. Wachenchauzer R. The Evolution of Computer Education in Latin America: The Case of Argentina. ACM Inroads. 2014;5(1):70–6.

232. Oreggioni I, Arbulo V, Castelao G et al. Building Up the national integrated health system. Uruguay case study. WHO. September 2015. www.who.int Accessed June 20, 2016

233. Ceibel Project http://www.ceibal.edu.uy/

Accessed May 15, 2016

234. One Laptop per Child Program http://one.laptop.org/ Accessed May 10, 2016

235. Text to Change Mobile www.ttcmobile.com Accessed May 5, 2016

236. Health Canada. Canada's Health Infrastructure. http://www.hc-sc.gc.ca/hcs-sss/eHealth-esante/infostructure/hist-eng.php Accessed January 24, 2016

237. Zelmer J, Hagens S. Advancing primary care use of EMRs in Canada. Health Reform Observer.10 October 2014. Volume 2. https://escarpmentpress.org/Accessed January 6, 2016

238. Infoway. The path of progress. Annual Report 2014-2015. https://www.infoway-inforoute.ca/en/component/edocman/2771-annual-report-2014-2015/view-document Accessed February 2, 2016

239. Canada Health Infoway. Health Policy Monitor. Torgenson R. October 10, 2006. http://www.hpm.org/ca/b8/2.pdfAccessed February 20, 2016

240. Office of the Auditor General of British Columbia. EHR in British Columbia. February 2010. https://www.bcauditor.com/sites/default/files/publications/2010/report_9/report/bcoag-electronic-health-records-ehr.pdfAccessed February 20, 2016

241. Canada Health Infoway. EHRS Blueprint: An interoperable EHR Framework. Version 2. March 2006 http://www.cas.mcmaster.ca/~yarmanmh/Recommended/EHRS-Blueprint.pdf Accessed February 20, 2016

242. Webster P. E-health progress still poor $2 billion and 14 years later. CMAJ. 2015. Doi:10.1503/cmaj.109-5088

243. Canada Health Infoway. First digital health week launches in Canada. November 10, 2014. http://www.infoway-inforoute.ca/en Accessed February 20, 2016

244. Canada Health Infoway. Leaver C, Hagens S. Impact of patients' online access to lab results in British Columbia. May 14, 2014. https://www.cahspr.ca/en/presentation/538486ba37dee8572cd5018f Accessed February 25, 2016

245. Mosher D. A framework for patient-centered care coordination. Healthcare Management Forum. 2014. http://hmf.sagepub.com/content/27/1_suppl/S37.full.pdf Accessed February 28, 2016

246. Mobile Health Computing Between Clinicians and Patients. Emerging Technology Series. White Paper. April 2014. http://www.infoway-inforoute.ca/en Accessed March 2, 2016

247. Samsung Canada and Saint Elizabeth team up to expand mobile innovation. 14 January, 2014. http://www.saintelizabeth.com/ Accessed March 3, 2016

248. Hernández-Ávila JE, Palacio-Mejía LS, Lara-Esqueda A, et al. Assessing the process of designing and implementing electronic health records in a statewide public health system: the case of Colima, Mexico. J Am Med Inform Assoc 2013;20(2):238–44

249. Pérez-Cuevas R, Doubova S V, Suarez-Ortega M, et al. Evaluating quality of care for patients with type 2 diabetes using electronic health record information in Mexico. BMC Med Inform Decis Mak BioMed Central; 2012 [cited 2016 Jun 24];12(1):50.

250. eHealth Strategy Tool Kit. https://www.itu.int/dms_pub/itu-d/opb/str/D-STR-E_HEALTH.05-2012-PDF-E.pdf Accessed May 2, 2106

251. International Medical Informatics Association. www.imia-medinfo.org Accessed May 3, 2016

252. WHO Atlas of eHealth Country Profiles. 2015. http://www.who.int/goe/publications/atlas_2015/en/ Accessed September 15, 2016

253. HL7 FHIR. https://www.hl7.org/fhir/ Accessed June 25, 2016

254. Millard PS, Bru J, Berger CA. Open-source point-of-care electronic medical records for use in resource-limited settings: systematic review and questionnaire surveys. BMJ Open. 2012;2(4)

255. Hersh W, Margolis A, Quiros F, Otero P. Building A Health Informatics Workforce In Developing Countries. Health Aff 2010;29(2):274–7

256. Luna D, Otero C, Marcelo A. Health Informatics in Developing Countries:

Systematic Review of Reviews. Yearb Med
Inf 2013;8:28-33

257. Luna D, Almerares A, Mayan JC, et al.
Health Informatics in Developing Countries:
Going beyond Pilot Practices to Sustainable
Implementations: A Review of the Current
Challenges. Healthc Inform Res World
Health Organization; 2014 ;20(1):3

258. William Gibson. Wikiquotes.
https://en.wikiquote.org/wiki/William_Gib
son Accessed May 25, 2016

INDEX

Lightning Source UK Ltd.
Milton Keynes UK
UKOW07f2344050917
308657UK00004B/245/P

9 781365 524806